Kids Design

童心童趣——儿童产品设计

马春东 编著

曲　义 译

大连理工大学出版社

Dalian University of Technology Press

图书在版编目 (CIP) 数据

童心童趣：儿童产品设计：英汉对照 / 马春东编著；曲义译．—大连：大连理工大学出版社，2013.8
ISBN 978-7-5611-7844-7

Ⅰ．①童… Ⅱ．①马… ②曲… Ⅲ．①儿童－工业产品－造型设计－世界－图集－汉、英 Ⅳ．① TB472-64

中国版本图书馆 CIP 数据核字 (2013) 第 103746 号

出版发行：大连理工大学出版社
（地址：大连市软件园路 80 号 邮编：116023）
印　　刷：利丰雅高印刷（深圳）有限公司
幅面尺寸：250mm × 215mm
印　　张：13.5
出版时间：2013 年 8 月第 1 版
印刷时间：2013 年 8 月第 1 次印刷
策划编辑：张　群
责任编辑：裘美倩
责任校对：王丹丹
装帧设计：张　群

ISBN 978-7-5611-7844-7
定　　价：98.00 元

电话：0411-84708842
传真：0411-84701466
邮购：0411-84703636
E-mail: designbooks_dutp@yahoo.com.cn
URL: http://www.dutp.cn

如有质量问题请联系出版中心：（0411）84709246 84709043

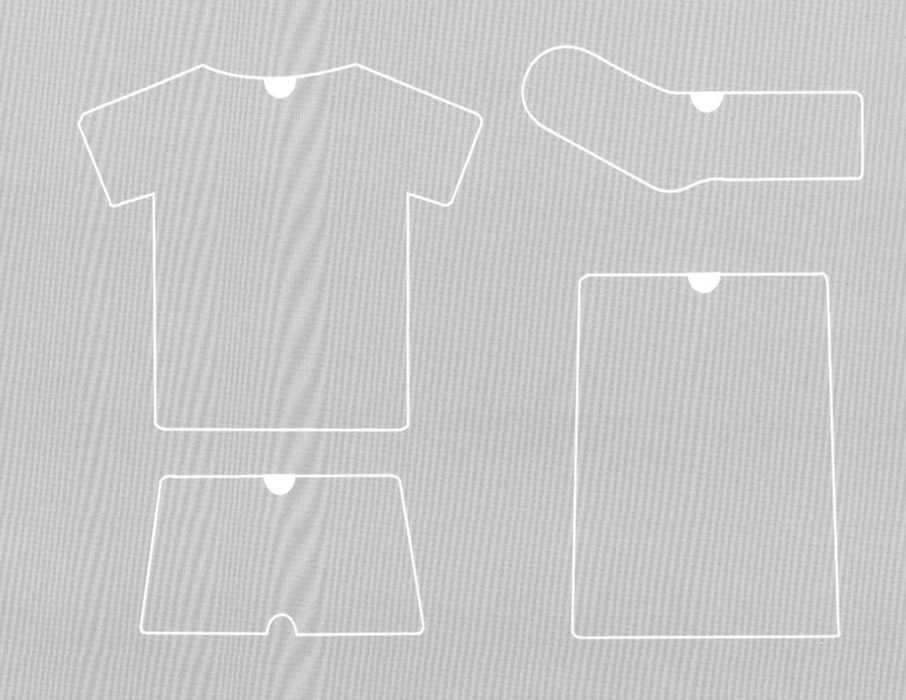

contents

目 录

为儿童而做的：与设计师或孩子看护者分享的话

人一生最重要的阶段要数 11（或 14）岁之前的成长，它对孩子以后的发展起着极其重要的作用。这是一个从生理发育、智力发展到心理健康乃至人文素质或价值观初步形成的各种成长因素相互作用同步发展的阶段，而且发展速度极快。孩子没费吹灰之力能在五六年的时间里学会跑跳玩耍、语言表达、察言观色、讨好耍脾气等等。大人们往往应接不暇，总感觉孩子不知不觉地突然长大了，于是乎便感慨起来：后悔当初没教孩子“这个”，没学“那个”，没研究儿童的成长规律，留下了很多很多无法弥补的遗憾……

保证孩子身体健康，避免伤害，使其正常成长当然是养育孩子基本的标准，亦即儿童产品的首要问题是要关注儿童的安全问题。儿童产品的安全问题不仅体现在其几何尺寸、结构强度、材料、重量等物理和化学指标方面，还应该考虑因使用方式和过程而影响到孩子的心理反应。一般人只习惯于从感官上关注孩子身体外在的变化，设计师应该考虑得更加周全。

关注孩子身心协同成长应是设计师遵循的重要设计原则。

身心协同成长指的是任何一种外界信息都会通过小孩的感官作用于神经、神经系统的发育与成熟；感知、认知水平的锻炼；意识和判断力的形成，反过来意识、感知水平又会促进神经等生理因素的健康成长，这就是大自然神奇魅力所在。小孩的成长不是机械式地等身体长好了，才开始学习。孩子的学习是身体生长的需要，是生理现象，所以，孩子的学习和进步能力是惊人的。可是往往因大人们的偏见，以成人的价值标准，以“小孩子不懂事”的观念，扼杀了孩子们伟大的智慧和阻碍了成长速度甚至造成了无法弥补的遗憾。一位美国儿童心理学家在直接养育十名不同家境的孩子达十七年之久后感慨道：“儿童的养育，不是教会他什么，而是别让他失去什么。”殊不知儿童是老师，大人才是学生。对小孩来说，任何器具都可能变成玩具，任何玩具都是教具，因此，儿童产品无一例外地都会直接作用于孩子的身心健康。设计师的责任重大而富于挑战。

一件优秀的儿童产品，必定是对应儿童年龄成长特点，绝不可能出现适合 1—5 岁如此大跨度的玩具。新生儿的成长变化是按天来计算的，幼儿是按月计算的，儿童是按半年或年计算的。孩子越小，变化时间越短，相应的产品生命周期就越短，由此，针对这时期的产品解决“浪费”问题便成为设计的重要指标。

……

据此，儿童产品一般的设计原则应符合以下几点：1. 安全——材料类别及材料有害物质含量（符合标准），使用过程中的易磕碰部位的处理，适龄的物理尺寸等。2. 适龄原则——尺寸，材料质地，色彩，功能，易适龄变化等。3. 营造娱乐氛围——营造适合年龄成长特点的心理环境，游戏氛围，吻合视、听、触等感官及认知能力发展要求。

为了宝宝身心同步健康成长，希望大人们多研究一点有关儿童成长的问题，且不应仅凭感觉草率地做判断。

此书选择了国内外一些较优秀的儿童产品案例，希望能够给读者带来启发。

Products for Children: words shared with designers and children carers

Before the age of 11(or 14), children would come to the most important life stage during which they are experiencing the maturity of body and mind, the formation of humanity quality and world values and the interaction of various problems.Within these few years, children usually make rapid progress like learning to run, to play, to speak, to observe, to please, and even to develop bad temper, while parents find it too fast to accept all the changes that happen to the little personalities and feel that they should have done better in their early education.

For parents, to meet the basic standard of raising kids is to provide them a healthy and safe growing environment. Thus, children's products also focus on the safety issue concerning the shape, structure, material, weight and more importantly, how they are used. While most people care about the physical change , the designers are more considerate in offering children the best gift for the development of their inner world.

A cardinal principle of all designers is to concern the harmonious development of kids' body and mind which means all information from outside world, through the children's sense organ, would act on the development of their nerve and nerve system and thus leading to the maturity of the perception, cognition, consciousness and judgment; in return, the higher level faculties would accelerate the healthy formation of other physiological elements. Accordingly, we come to know that the development of a child doesn't start from the maturity of their body, and learning is also part of the development. But most parents fail to realize the striking capability of their kids' learning and believe that they are too young to understand things. As a result, their judgment strangle the kids' great wisdom and hamper their growing speed. An American child psychologist who once personally brought up ten children from different family circumstances for 17 years says that the key point of children raising is not how much you have taught them but how they are taught to avoid losing. In this sense, children are not students but teachers and they may turn every utensil into toy and every toy into a teaching tool. Therefore, all children's products will, without exception, act on the sound development of children's body and mind. In this sense, designers are specially expected to meet the great challenge and fulfill the parents' requirement.

An excellent work is designed to cater to children's age characteristic and not to satisfy the needs of children with large age span, e.g. from 1 to 5years old. Usually, the change of a new born baby is based on days, infant on months and kids on years or half a year, thus the product for younger children is characterized with short life circle, a feature must be considered in order to design practical pieces of work.

Based on the above mentioned factors, the children products are designed on the following principles as :

Safety: The material and the content of harmful substances confirm to the standard; the edges and corners are well treated; the dimension is for proper age.

Right age: The size, material, color and function of the products are for right age.

Entertainment scene: To create proper mental environment and mood of game, appealing to the senses of sight, hearing and touch, in line with the cognitive ability.

The adults are thus expected to pay more attention to the children's development of mind and body instead of jumping to conclusion with feeling.

This book selects some excellent children products from foreign countries and hopefully, you will be inspired by these good examples.

睡　的　设　计

Products for Sleeping

儿童的第一件家具要数床了，新生儿一天要睡 16 个小时， 2—3 岁要睡 12 个小时，也就是说，婴幼儿一天的一半时间都处在睡眠之中。睡眠最能使大脑得到休息并使其健康成长。新生儿和哺乳期婴儿异相睡眠占 40%—50%，而成人只占 20%，可见让婴幼儿有个好的睡眠是多么重要。因而，严格遵循儿童床的设计标准是非常重要的。

儿童身体变化很快，床的几何尺寸亦应随之改变，这对设计师来讲是一个非常头疼的难题。婴幼儿时床面的大小(较宽松为宜)、围栏的间隙 (一般不超过 6 厘米，防止儿童头卡住)、高度 (50 厘米为宜，过低宝宝容易翻出，过高大人抱放不便)、床体的高度、褥子的厚薄 (5—10 厘米为宜)、软硬透气度及大小 (距床边的缝隙不超过 4 厘米) 等要求比较苛刻。儿童各年龄段身体特点差异很大，若想解决这些问题，同一款式的床不宜用得太久，或者说很难做到同一款床适合 1—8 岁的儿童，由此，儿童床的使用寿命便成了设计的难题。设计师们为了解决这个问题，探索了各种各样一次性、易拆易装或改变功能的组合方式，以求最大化地节约和利用。

儿童床围护栏若采用绳索之类材料，要充分考虑结节的间距，以免缠绕脖子发生窒息。一般来说，绳状物的东西不宜在床内留放。

儿童在长牙时喜欢啃咬，床的表面涂装一定要防止油漆龟裂和含铅、苯、甲醛等有害物质。

吊篮、摇篮床的摆动频率和幅度以母亲怀孕时行走的频率和幅度最为适宜；摇床的设计要特别注意震动的强度，过分的摇晃会使婴儿的大脑在颅骨腔内受到震动，轻者影响脑部的生长，重则会使尚未发育成熟的大脑与较硬的颅骨相撞，造成颅内出血，这对 10 个月内的孩子尤为危险。

儿童床稳定性要好，摇篮床内不要有突出物，定期检查床的稳固程度，表面要光滑，上下两层床要充分考虑儿童的游玩方式和年龄特征等等。

新生儿初生几天后大部分时间是处于睡眠状态。这是因为其大脑皮质还不能适应外界刺激的强度。虽然对一般人来说是一些普通的刺激，而对儿童则是超强的刺激，这种超强的刺激就引起保护性的意识——睡眠。因此，保证婴幼儿的睡眠是至关重要的。

总之，儿童床重要的设计、制造和使用原则要充分考虑儿童的年龄特征和环保标准，不能仅从成人的审美视角来判断。

Bed is the first household furniture for children since newborns sleep for up to 16 hours a day during the first few months and those of 2or 3 years old will sleep 12 hours a day. That is to say, babies spent almost half a day sleeping which is helpful in developing a healthy mind. The paradoxical sleep periods of newborns and nursing babies accounts for 40% to 50%, while that of the adults only 20%, so a good sleep is essential for babies and thus the design standard of their beds is also specially important.

Babies grow fast and their beds are expected to grow with them. But it is a challenge for designers to fulfill the expectation since there are some basic standards to have to be followed in producing the bed, such as the width of bed(the wider the better), rail gap's width (not exceeding 6 cm) and its height(50cm is favored), the height of bed, the thickness of the mattress(5cm to 10cm is favored), its air permeability and size (a gap not exceeding 4cm from the bed). Therefore, it is hard to create a single bed that can enable use from new born up to 8 years old concerning the facts that children will experience dramatic physical change during these years. Accordingly, designers are encouraged to explore different types of combination which are disposable, easy to install, disassemble or assemble, in order to maximize the beds' function and value.

If ropes (not supposed to be put on bed) are used as guard rails, special attention should be paid to the gap between knots to protect the babies from stifling.

Avoid paint cracking and toxic substances (like lead, benzene, or formaldehyde) in coating the beds since babies are apt to bite when teething.

The swing frequency and angle of cradles or cribs are expected to correspond to those of pregnant mothers. The vibration of cribs should be managed to avoid the potential danger aroused by excessive shake: at best, it will result in the slow brain growth; at worst, an intracranial hemorrhage. Give special care to 10-month-olds or younger.

Check the stability of little beds regularly; keep the beds clean and the surface smooth; consider the playing mode and age when designing the upper and lower beds.

Offer babies the best sleeping environment since newborns spent most of their time sleeping, a protective inhibition adapting to the outside stimulus.

In brief, age characteristics and environmental standard, not the adults' aesthetic perspective, is the top priority in designing, producing and using the children's beds.

Knoppa

Knoppa
W85cm×D40cm

这是一款适合婴儿期使用的摇篮，可上下自由调整、拆解、打包，方便旅行时使用。布包可机洗。

Knoppa is a cradle that your child can use from birth. The cradle can very swiftly and easily be raised or lowered, disassembled, packed and brought with you when you travel. The cloth bag can be machine washed.

Uffizi Bunk Bed

argington

The Uffizi Bunk Bed 为充分利用孩子的房间提供了最优方案，更为孩子创造了属于自己的休息空间。手工匠 duo Jenny Argie 和 Andrew Thornton 设计的 Uffizi 搭配趣味性空间，堪称一部建筑作品，孩子们一定会爱上悬臂、半封闭下铺以及类似开口的窗户。顾客还可以为床的不同部分选择不同的外观饰面效果，形成个性化风格。Argington Fuji 玩具箱是位于床脚的绝佳附件，可满足额外的储存需要。CPSC(消费品安全委员会) 建议上床适用于 6 岁以上儿童。

The Uffizi bunk bed offers a straight forward solution to making the most of your children's room. More importantly, it provides a resting place your children will love to call their own. Designed by the artisan duo Jenny Argie and Andrew Thornton, the Uffizi can be considered as an architectural work given the interesting spaces it provides. Kids will love the cantilevers, semi-enclosed lower bunk, and window like openings. You may select different finishes for different parts of the bed giving you the opportunity to suit your own style. The Argington Fuji toy box would also be an excellent addition at the foot of the bed to provide additional storage needs.The CPSC recommends that the upper bunk should be used by children ages 6 and up.

Nina's House

Knoppa
D124cm×W111cm×H170cm

房中房，屋中屋！ Nina's House 专为新生儿创造了一个独立空间，使家长免于对房间进行翻新和移动。

这个小房子能满足孩子的所有需求：婴儿床（婴儿护栏）、存储柜和更换尿布的空间。只需用膝盖轻轻一推就能轻松打开和关闭抽屉，以便家长腾出双手为孩子更换尿布或衣服。

有趣的切割窗口确保父母在孩子睡觉或玩耍时也能随时注意他们的动静。整个房顶可被向后掀开卷起，由明亮的黄色带子连接，十分灵活。孩子长大后，这款屋子仍然适用，降低围栏，去除护栏，就能再造一个新床，下方的两个轮子可以使其自如移动。另外，房子也可以被拆卸压扁，方便运输和储存。

A house within a house, a room within a room. Nina's House was created out of the need for a separate space for a newborn, without having to renovate or move house entirely.

The little house accommodates all the necessities for a baby: a crib/playpen, storage and a place to change diapers. The chest of drawers below can be opened and closed with a slight push of the knee so that both hands can still be available while changing the baby's diapers or clothes.

Playful cut-out windows ensure that parents can still keep an eye on the little one while he or she is playing in the playpen or sleeping. The entire roof can also be opened up by rolling back the roof tiles which are connected flexibly with bright yellow ropes. The house can be adapted as the baby grows older. To create a bed, the base of the playpen can be lowered and the rails can be removed. Underneath, the house rests on two wheels allowing the house to be moved around easily. Otherwise it can be disassembled back into a flatpack for easy transport or storage.

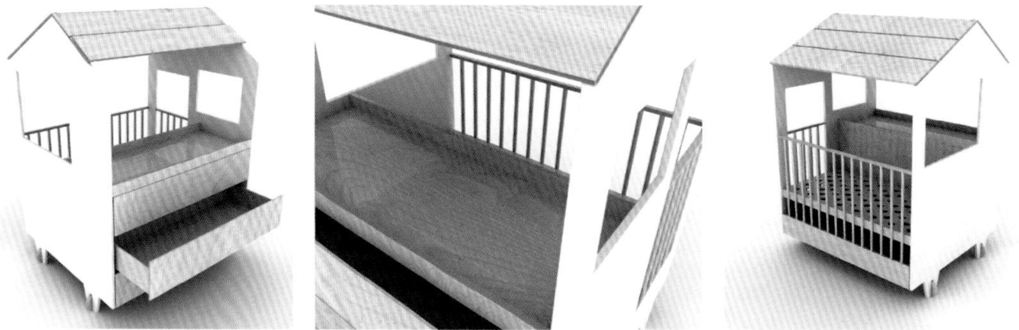

Alma mini™ urban crib

Bloom

有两层床垫高度的 alma mini™ 结合了新生儿摇篮与独立婴儿床的优势，使父母可以轻松接触到宝贝儿，并为孩子留足了成长空间。

无需任何工具的无障碍安装，与可锁蓖麻车轮相辅相成，可轻松移动，适合折叠储藏安置。挡板四面都有开口，充分保证孩子成长所需的空气流通，可供新生儿至两岁左右儿童使用。

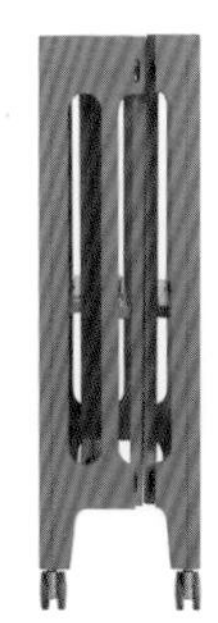

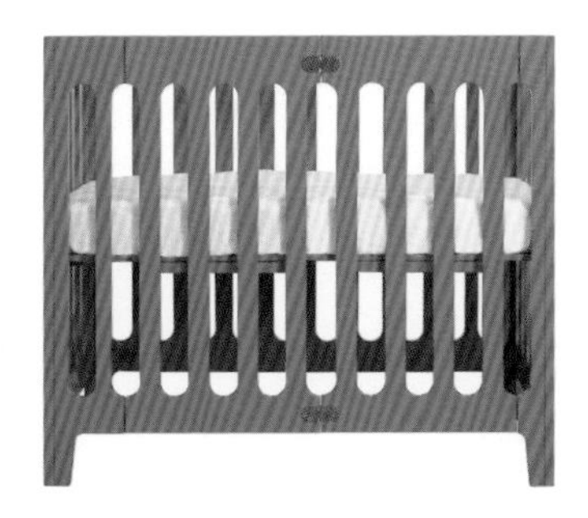

alma mini™ has two mattress heights combining the benefits of a new born bassinet or cradle with a cot/crib; parents have easy access to baby and there is room for baby to grow.

No-tools-required hassle-free assembly is complemented by lockable castor-wheels for effortless mobility and the signature stow-away fold. alma mini™ open slats on all four sides maximize all-important airflow for baby. A modern easy folding bed that can be used from newborn to two years.

Bloom coco™ stylewood™

Bloom

cocoTM stylewood™ 婴儿躺椅由压制木材制成，采用五点安全无障碍设计，软脚座椅，可以进行舒适平稳的摇摆运动，使孩子能够按照自己喜爱的姿势休息。本婴儿躺椅适合新生儿及体重在 12 千克以下的婴儿使用，无需工具即可安装。安全性是 bloom 公司一贯的考虑重点，coco 座位采用了五点可调节的安全设计。

The coco™ stylewood™ baby lounger is crafted in pressed wood, it enables a smooth rocking motion in the comforts of an angled soft-feel seat with 5-point safety harness allowing babies to rest in style.

coco™ stylewood™ baby lounger is suitable for babies from newborn to 26 lbs. with a design of no-tools-required assembly.safety is always one of the most important concerns from bloom, coco seat pads feature adjustable 5-point padded safety harness.

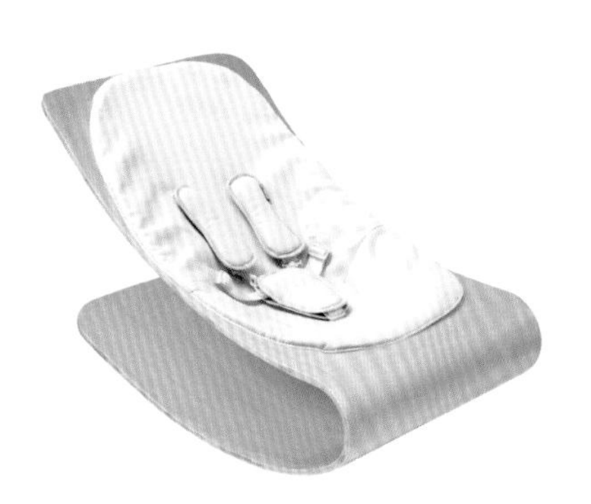

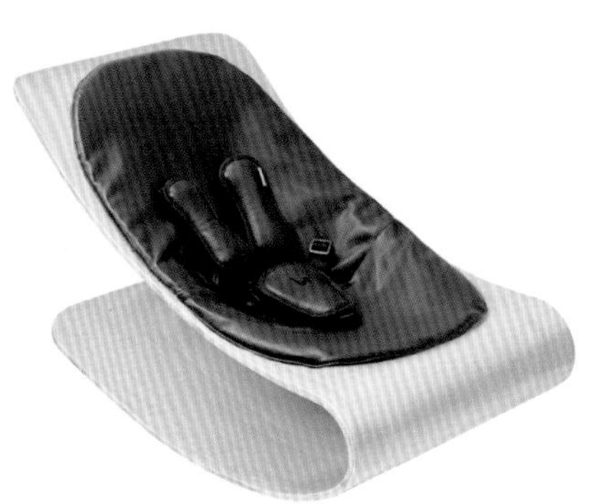

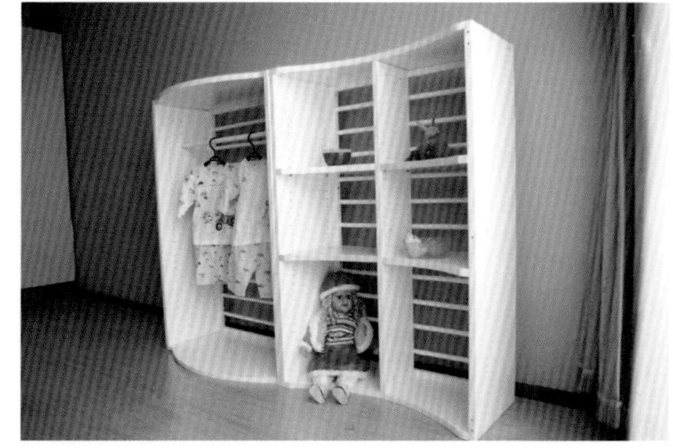

Convertible Multifunctional Bed 2011G

www.asai-hagou.com

2011G 型号床设计理念超前，实用性强，外观简洁。前期作为婴儿床使用; 重新组装可以变换成儿童衣柜，书架。外观设计简洁，创意性和使用性强；组装、变换方法简单。

2011G is innovative and practical with simple appearance. It can be used as a crib and later a wardrobe or bookshelf for an older one. The design features simplicity, originality, utility and easy assembling and conversion.

Sahara Crib

argington
D57cm×W33cm×H40cm

Sahara 婴儿床物美价廉，时尚现代。改装为学步车后可提供更多功能，发挥更大作用。Sahara 多屉橱柜与任意可变托盘的搭配使用使育婴室更加完美。

The Sahara crib is a sleek, contemporary crib at an affordable price. It converts to a toddler bed giving parents extended use and a great value. It is the perfect match with the Sahara Chest of drawers and the Anywhere Changing Tray to completely outfit your nursery.

Cradles Boris, Vladimir, Alexander

nikazupanc

Ayres Twin Bed™

argington

该床使用实心桦木、低VOC含量的桦木板、涂料等可持续混合材料，将自然材料与周围环境完美结合，令房间散发现代气息。随床附件有可拆卸护栏以及活动床，活动床不会占用额外空间，适合临时过夜或小型公寓使用。另外，活动床、双人床以及护栏可以互换，能自定义多种组合。

With the use of mixed sustainable materials including solid birch wood and low voc birch plywood and finishes, the bed imparts the richness of natural materials to its surrounding environment giving any room a great modern look. Accessories offered with the bed include detachable guard rails and a rolling trundle bed. The trundle bed is a great accessory for the occasional sleep over or for small apartments that need extra beds without taking up extra space. Finally, the finishes for the trundle bed, twin bed and guard rails are all interchangeable, so you can customize the bed in a plethora of combinations.

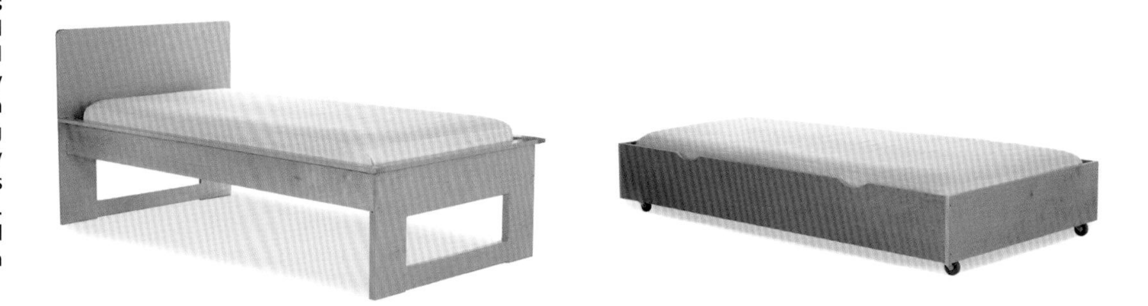

Fresco Loft™ Contemporary Baby Chair

Bloom

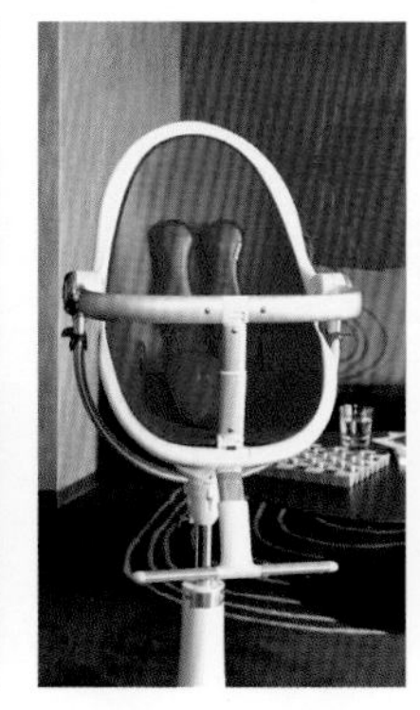

“三合一”位置模式伴随孩子共同成长：斜倚座位可以做新生儿摇篮用（新生儿至6个月休息摇篮模式），之后，根据孩子的成长需要调整座位高度（6个月以后喂食椅，可采用传统托盘喂食模式）。该座椅同时有可调节的搁脚板以增加舒适度和稳定性。

Fresco Loft™ features a 3 in 1 grow-with-child position mode: reclining seat that can be laid back to cradle newborns (newborn to 6-month resting cradle mode), then gradually brought upright as baby develops (6-month & up feeding chair with "up-to-the-table" & traditional (tray) modes). It has an adjustable footrest for added comfort and stability.

Koo-di Pop Up

W|H|L
L60cm×W30cm×H18cm

获奖作品 Koo–di Pop Up 是家长带婴儿外出旅行的首选。这款不足 3 公斤的折叠包小巧轻便（体积为 60 × 30 × 18cm），便于手提或直接装进行李。装有气垫和蚊帐的折叠包可在几秒钟之内打开，折成一个体积为 100 × 60 × 69 的大气垫床，能在任何地方为婴儿提供洁净安全的睡眠场所。

The award winning Koo-di Pop Up Travel Bubble Cot, was the first travel cot really suitable for travelling. Weighing only under 3kg, a very small 60 x 30 x 18cm, it can be easily hand carried or fit into your luggage when travelling. Complete with mattress and mosquito net, the Koo-di Pop Up Travel Bubble Cot opens in seconds to a very large folded yet opening to a very large 100 x 60 x 69 providing a clean, safe area for baby to sleep anywhere.

M147&M126 Purflo

W|H|L
L60cm×W30cm×H18cm

与传统泡沫气垫不同，Purflo 通气、耐洗，保证婴儿睡好，家长安心。

作为全球唯一可充分洗涤床垫，Purflo 为婴儿提供了真正防尘螨的洁净空间，有效防止哮喘和湿疹等会加剧孩子过敏的过敏源积聚。

The Purflo mattress replaces the conventional foam mattress with a fully permeable/ breathable, fully washable, foam-free mattress. An essential product to help parents minds be at peace while their baby sleeps.

The only fully washable cot mattress in the world, the Purflo mattress provides a genuine dust-mite free environment for babies, helping to prevent the harmful build-up of allergens which are known to aggravate childhood allergies such as Asthma and Eczema.

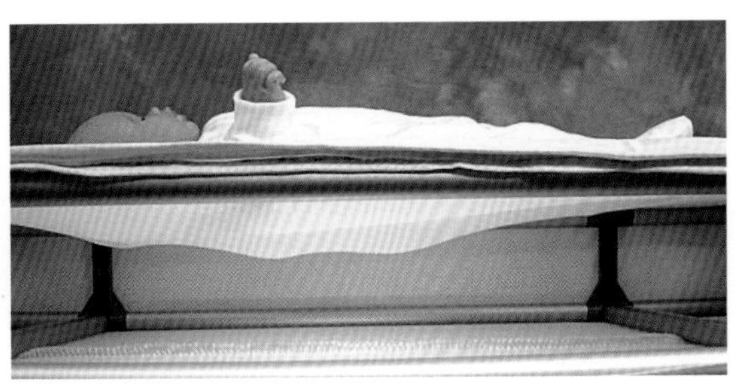

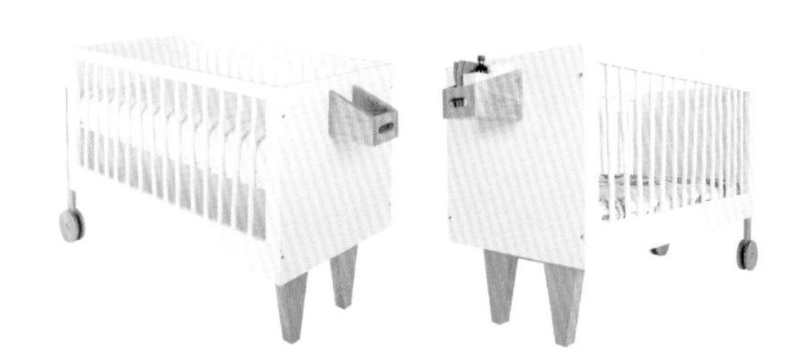

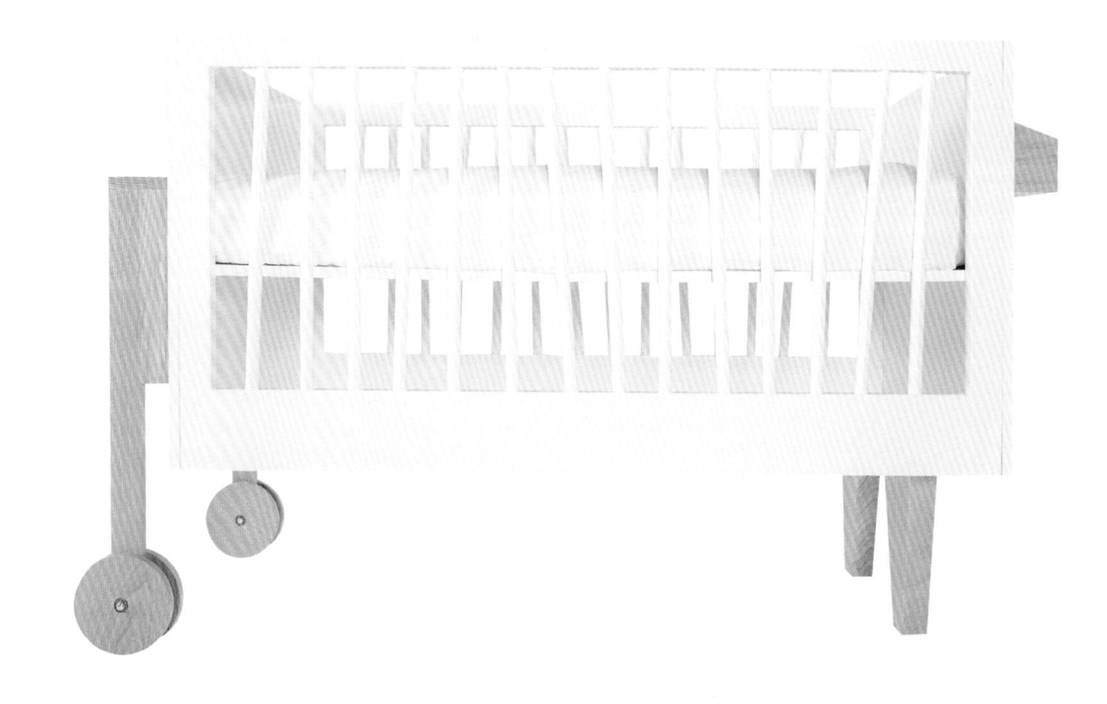

Zebra_02

ninetonine
Alberto Marcos for ninetonine
L158.5cm×W65.5cm×H96cm

现代 Zebra 的“头部”可轻松移动，亦可作为放置饰物和酒品的储存空间使用；“尾部”是一个实用性很强的搁板。Zebra 亦可变身为坐卧两用的长沙发，底部有一个高度可调的两层床垫。

Easy to move thanks to its "head", which is also a useful storage space for dummies and bottles, our modern Zebra_02 also has further practical shelving at the back in its "tail". Easy conversion into a daybed-sofa. Adjustable 2 height mattress base.

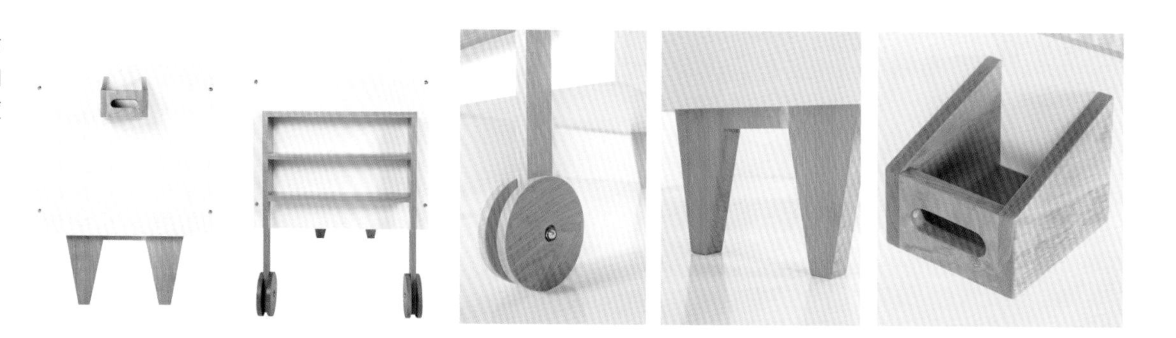

Neo Cotbed

ninetonine
Diezmasdiez for ninetonine
L150cm×W75cm×H100cm

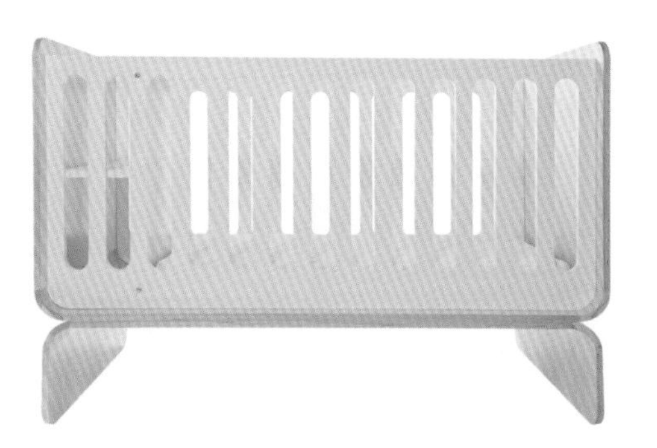

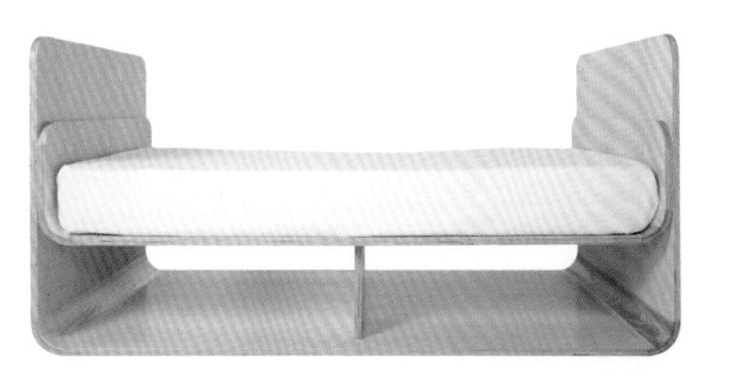

NEO 简易床中的可调节床垫非常舒适。超级实用的储存空间适合放置玩具、枕头和酒瓶。只需简单变换，就可轻松升级为一个现代、舒适的儿童床，床下储存空间宽敞。本套系不包括床垫。

Our great NEO cot-bed starts life as a cot, with an adjustable mattress base. Note its highly practical storage area, perfect for toys, pillows or bottles. It is then easily converted to a modern and comfortable junior bed, with ample storage space below. Mattress not included.

alma papa™ Crib

Bloom

alma papa™ 同样彰显了 Bloom 公司的特色——一个简单的折叠专利设计。可快速组装，无需使用工具，不用时便于无障碍折叠储存。配合可锁住的蓖麻车轮使用，使移动更加轻松自如。alma papa™ 四面均可打开的板材为婴儿提供最充分的空气流通。与之前的流行趋势一样，alma papa™ 专为城市生活设计，专注提高生活水平和睡眠质量，更适合祖父母家中或第二居所摆设使用。

alma papa™ 有两层床垫的高度，结合了新生儿摇篮与全尺寸婴儿床的优势，将其转变为一个令人惊叹的学步床，从而延长了 papa 在现代家庭中的使用寿命，适用于新生儿阶段至 4 岁孩童。

alma papa™ also boasts bloom's signature, a simple patented folding design, allowing rapid assembly and easy folding for storage when not in use with the no-tools-required hassle-free assembly. alma papa™ is complemented by lockable castor-wheels for effortless mobility and the signature stow-away fold. alma papa™ open slats on all four sides maximize all-important air-flow for baby. Like its popular predecessor, the alma papa™ is designed for urban spaces, elevated living, co-sleeping and is perfect for the grandparents or a second home. The alma papa™ also comes with 4 optional lockable castor-wheels.

alma papa™ features two mattress heights, combining the benefits of a newborn bassinet or cradle with a full size crib the alma papa™ bed rail converts papa into a stunning toddler/day bed, extending the life of papa in the modern home; enabling use from new born comfortably up to 4 years.

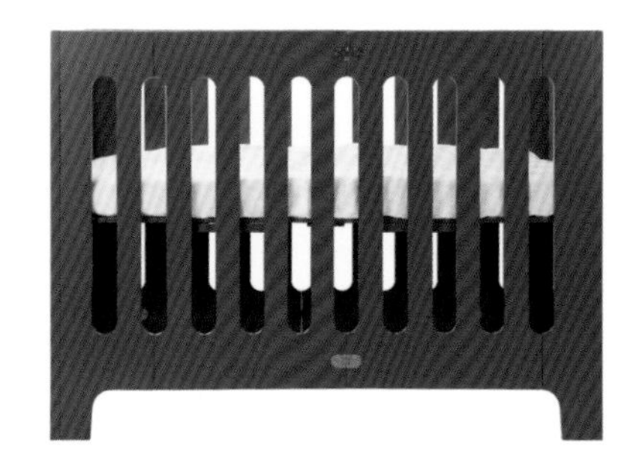

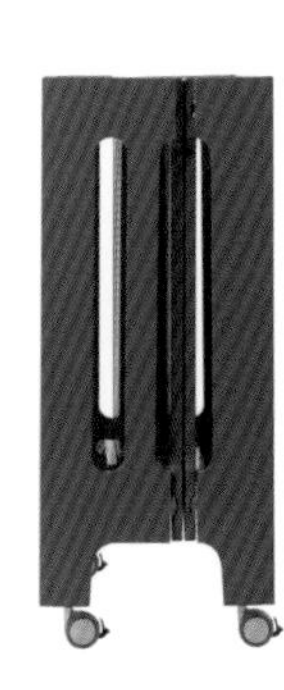

Duet Bunk Bed

Nurseryworks

Duet 双层床的下床底架有两种设计可供选择：三个宽敞抽屉或配备订制床垫的带脚轮矮床。另有额外储存空间，包括 6 个开放小搁架、下床橱柜和上床橱柜。产品由持久耐用的楸木及其他无毒材料打造而成，各项指标均达到最高检测标准，为孩子的健康睡眠提供舒适安全的保障。

非凡材质打造活力与简约，完美细节充分满足家人的视觉审美。

Select between two under-bed options: three spacious drawers, or roll-out trundle bed with custom mattress. Store more with six open cubbies, one lower bunk cabinet, and one upper bunk cabinet (safely accessible from upper bunk). Sleep better knowing the bed your children rest on is crafted from durable Catalpa hardwood and non-toxic materials, and tested to the highest standards.

To satisfy your family's visually-informed aesthetic, Duet harmonizes youthful, modern simplicity with unique details and unusual materials. Adults and kids alike will appreciate Duet Bunk Bed's tasteful, distinctive style.

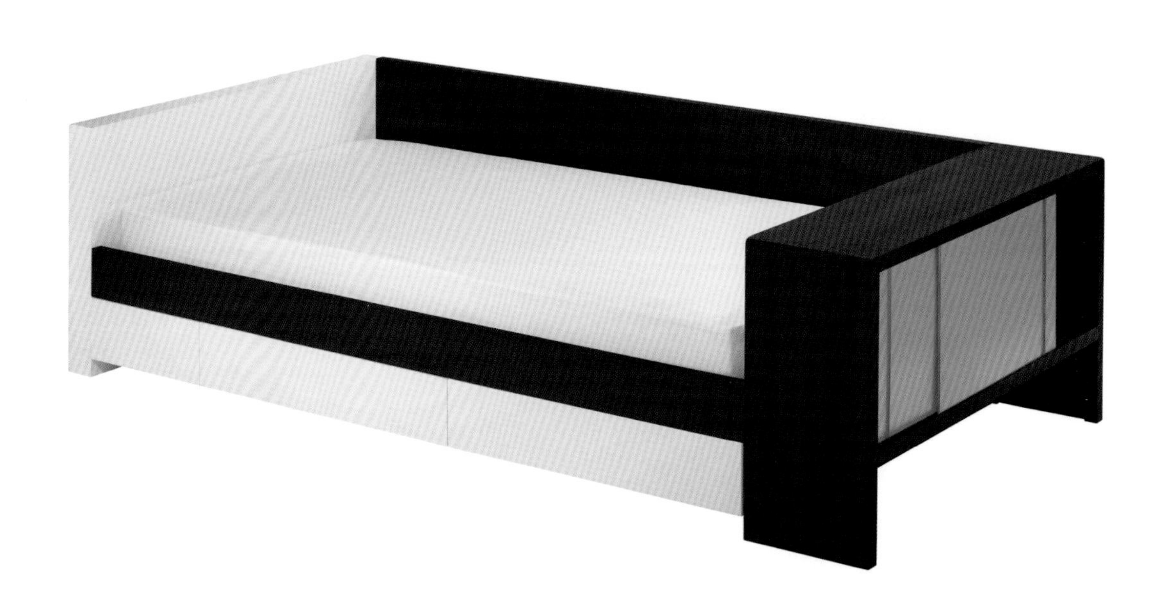

Duet Twin Bed

Nurseryworks

Duet Twin Bed 的活泼风格是孩童的至爱，而其现代风格与非凡材料亦会满足成年人的视觉审美。两种款式可供选择：一种设计为床下三个抽屉用于置放寝具、书籍或玩具。另一种为通宵派对的小朋友量身打造，可铺开矮床，配有订制床垫。

Duet Twin Bed's youthful vibe appeals to kids, while its contemporary form and unusual materials delight a visually-informed adult aesthetic. Select one of two twin bed options: three under-bed drawers to keep bedding, books and toys nearby, or a roll-out trundle bed with custom mattress for hassle-free sleepovers.

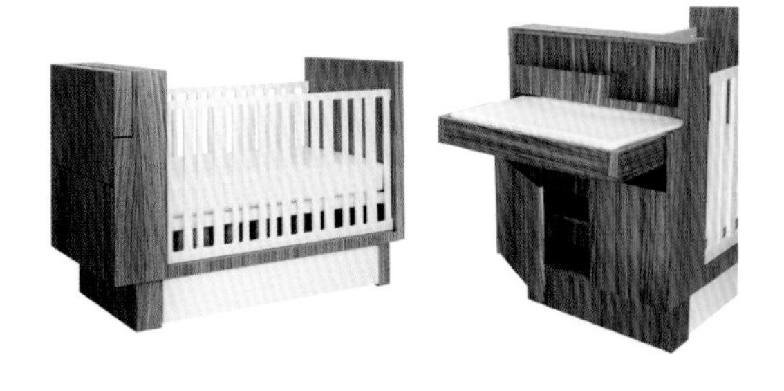

Studio Crib

Nurseryworks
L165cm x W77cm x H107cm

这是一款设计简便、安全的婴儿床，可充分满足育婴需求，材质坚固，质量可靠，确保宝宝的高质量睡眠。滑顺灵活的床下抽屉方便储存毛毯等寝具，可升降伸缩桌和宽敞隐式橱柜提供可变换的额外空间。

孩子长大后，可变桌能方便改装为嵌入式课桌和沙发床（坐卧两用床）或改装成配有护栏的幼儿床。

Studio Crib satisfies four nursery needs in one safe, easy-to-use design. Your baby will sleep peacefully in this sturdy and reliable crib. A smooth under-crib drawer conveniently stows blankets and bedding. For changing needs, Studio Crib's gently-lowering changing table (pad included) and spacious hideaway cabinet are easily accessed.

As baby grows to a toddler, the changing table easily converts to a built-in desk and the crib to a daybed. Conversion kit is also available to transition to a toddler bed with matching guardrail. Studio Crib's versatility provides your child with a safe haven for years.

Loom Crib

Nurseryworks
L141cm x W78cm x H95cm

Loom Crib 以其清洁、现代的俏皮风格重释了 Shaker Crib 的设计，其前瞻性育婴灵感来源于经典传统工艺。

护栏板条分布科学、制作精巧，不仅赏心悦目，更能保证婴儿安全，三面板条框架完美搭配雪白色嵌板。孩子学会爬行后，只需简单改装，Loom Crib 便可成为宝宝的幼儿床（改装组件独立包装）。

Clean, contemporary style leaves a big impression with this playful reinterpretation of the Shaker crib. The Loom Crib is a forward-thinking nursery jewel inspired by past craftsmanship.

Varied slat distances and thicknesses intrigue the eye, measured and crafted not only to delight but to ensure your baby's safety. Slats and frame are available in three finishes, all beautiful complements to the snow white panels. The Loom Crib accommodates baby's toddler years too, easily transitioning to a modern toddler bed once your child is climbing (conversion kit available separately).

Uptown Twin Bed

Nurseryworks

Uptown Twin Bed 融合了经典设计灵感与现代审美理念。其活力与文雅设计不仅契合清新精致的居家风格，也会令儿童房魅力倍增，是所有孩子的理想选择。

The Uptown Twin Bed plays with classic inspirations and bold, contemporary aesthetics. The youth and refinement of this piece add charm to your child's room, while suiting a clean, sophisticated home. The Uptown Twin Bed is ideal for both boys and girls.

Vetro Crib

Nurseryworks

作为市场上独一无二的全亚克力制造婴儿床，Vetro 婴儿床延续了 Nurseryworks 在儿童产品设计上的一贯先锋优势与魅力，为孩子营造优质睡眠的舒适环境，与孩子共同分享创造力与创新力。透明床板方便家长护理，并可与四周任何色彩搭配。专利设计包括优美的弧形边缘和三档可调节高度床垫。

Introducing the Vetro Crib, a one-of-a-kind acrylic crib. Nurseryworks continues its legacy as a pioneer in children's design with this swoon-worthy, entirely acrylic crib, the first and only in the market. Celebrate creativity and innovation with your beautiful little one. Your baby will sleep peacefully in unparalleled contemporary style and supreme safety. Clear sides give parents an unimpeded view of their child, and suit all color schemes. Our patented design includes graceful curved edges and three adjustable mattress levels.

Baby Diamond Bed

Atelier Charivari

罕见的两条 20 世纪 30 年代灰色阴影涂装。

Very rare bed from the 1930's painted in two shades of grey.

Kid 1960's Bed

Atelier Charivari

20 世纪 60 年代风格的浅灰色木质床。

A wooden bed from the 1960's painted in a light grey.

Baby Crib

Atelier Charivari

20 世纪 50 年代婴儿床的样式。

1950's Crib on heels.

Convertible Multifunctional Bed 2011D

ASAI

可供婴儿、儿童、大人三阶段变换，长期使用，实用性强：随床赠送三折床垫，用作婴儿床高度调整，以及儿童，大人床普通床垫使用（大人床需加长部件）；组装，变换方法简单，女性一人即可操作；可承受大人体重强度。

The practical and convertible bed can be used by new born up to adult with three mattress adjusted for a growing one of different stages(extended parts are needed for adult bed). Easy assembling and conversion, operated by one person, solid enough to bear the weight of an adult.

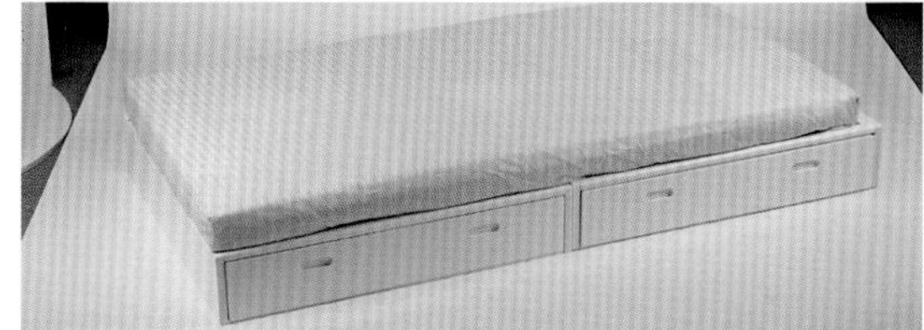

Convertible Multifunctional Bed 2011F

ASAI

从婴儿到大人都可以使用的可变换型床：从婴儿床可以变换成大人床，强调多功能性，实用性；床身自带抽屉，为妈妈提供便利的储物空间；组装和变换方法简单，女性一人可以操作；为确保儿童安全，我们选择了开闭式门设计；可以承受住大人体重强度。

The piece of work can be easily converted and used for both baby and adult with storage drawers. Easy assembling and conversion, operated by one person, solid enough to bear the weight of an adult. Door of open-and-close type is designed to protect your baby.

Bam Bam Crib™

argington
L139mm × W76mm × H85mm

巴姆婴儿床套件包含巴姆巴姆摇篮、巴姆巴姆婴儿床改装部件、巴姆巴姆婴儿学步床组装部件、一块泡沫床垫以及带有机棉外罩的摇篮。

床垫可以调节为三种高度，婴儿床便于组装、结构紧密、床体轻、设计高效。达到并超出 ASTM、CPSC，以及加拿大卫生部提出的所有标准。

BAM BAM crib complete includes the BAM BAM bassinet, BAM BAM crib conversion kit and BAM BAM toddler conversion kit. The set comes with a 1" foam mattress pad and organic cotton quilted mattress cover for the bassinet.

Crib Mattress is adjustable to three positions. Crib is easy to assemble, compact, lightweight and designed with efficiency in mind. Conforms to and exceeds all regulations set forth by ASTM, CPSC, and Health Canada.

HANNAH ♡'s
Daddy

Smart Kid

Adensen Furniture Ltd.

多功能家具组合 Smart Kid 像一间盒式卧室，使用年限长，是家长梦寐以求的理想选择。

设计简易的小床侧面和底面分别设有储物抽屉，顶部配有伸缩桌，只需简单调整，就可使孩子们体验不同的家具类型，拥有一个大型儿童床。本款组合家具的设计理念融入伸缩原理，让孩子的房间更简洁、紧凑，改装成游乐室后可提供更宽敞的生活空间。

本套系使用年限长，经济环保，床具采用上等桦木与胶合板，外涂无毒性漆，6 种颜色可供选择。我们相信家具的天然材质与简洁造型能使置身其中的人们身心平和——简约即美。

The multi-functional furniture Smart Kid is like a bedroom in a box and rightly something that every parent has dreamed of. It has designed to last for years.

It starts its life as a compact cot, complete with handy storage drawers at the side and at the bottom plus a changing table on a top. This furniture set allows the kid to have various types of furniture with only slight adjustment and the cot transforms into a bigger child's bed. The thoughtful design of the particular modern furniture keeps the child's room to have neat and compact appearance by using some retractable mechanism. More living space becomes available if you switch the bed furniture into a playroom mode.

This furniture set is ecological as well as economical, because it can be easily used for many years. The kids furniture is made of a high-quality birch solid wood and a plywood and finish by non-toxic water-based lacquer. There are 6 basic colors available.

The designer believes that the people who are surrounded by natural materials and laconic shapes become peaceful – they start remarking the beauty of simple things.

Leander Bed

Leander

孩子们转眼间就会长大成人，所以尽情享受与小家伙们在一起的欢乐时光吧！

Leander 儿童床的独到设计追随孩子的成长脚步，见证他们的成长历程（婴儿期到孩童期）。本套家具五件合一，各部分重新组装后可满足孩子不同成长阶段的需要。包括床垫在内的各部分构件买床时全部附带，床体造型圆滑、柔和，弹簧床垫透气性好，令睡眠舒适惬意。

Leander 儿童床可根据孩子的年龄增长做相应调整，免于环境变化带来的不适，独特的无棱角椭圆造型给人柔软温和之感，为幼儿带来贴心的安全感。

本设计集简约性、功能性、审美性、环保性与耐用性为一体，为您精心打造高品质生活。

儿童床的特殊造型彰显其天然气质与可辨的整体性，“万变不离其中”，就像不断成长变化的宝贝儿一样。

Enjoy the moment while the child is a new little person because they grow up so quickly. The Leander bed changes with the child and follows the child's development; from the baby age to the junior age. The bed is five pieces of furniture in one design. It is built in parts that are assembled so the bed always suits the child's age and needs. You get all parts including mattress parts when you purchase the bed. The round, soft forms make the Leander bed a wonderful piece of furniture to use. The spring mattress contributes not only to a comfortable night's sleep but also ensures good ventilation.

The Leander bed follows the child so he or she does not experience those changes in the environment that can seem insecure in the world of a little child.

The bed is constantly the same and can be adapted to the child's stage in life. I have retained the soft, oval expression and have deliberately designed the elements so they are not an angular system of blocks.

The designer has tried to combine the simple, the functional and the aesthetic – and the materials had to be friendly and solid. The designer wanted to create a piece of furniture that delights and raises the quality of life.

The special shape also means that the bed is a natural and, not least, recognisable entity at all stages of development. Even though the bed changes and develops, it is still the same. Just like your child.

Leander Cradle

Leander

本款摇篮高于地面，可通过四根带子系于一点悬浮在天花板下或挂在专用支架上。

婴儿在悬空的摇篮中自由晃动，如同在妈咪腹中一样舒适，而当他们“意识到”自己能轻易晃动摇篮时，定会平添无限乐趣。

自由晃动的摇篮还可刺激孩子方向感的形成，这一感觉神经元位于内耳，不仅有助于身体达成平衡，还可以帮助感知身体与周围事物、方向以及速度的相对位置。例如，感觉神经元会“指挥”我们在抓球时如何站位，帮助大脑在周围的复杂情形中选择目标。

方向感的健康发展还有利于肌肉平衡感的获得，反之，孩子可能很难在椅子上坐稳并集中精力学习。

The Leander cradle is a cradle that does not stand on the floor. The designer has created it so it can be suspended from the ceiling. Four straps are gathered into one point, either in a hook in the ceiling or in a stand made especially for the cradle.

Suspending the cradle meant that the cradle does not just swing in one direction and the movements become calm and free so they resemble the movements the child experienced in his or her mother's tummy. At the same time, the cradle is easy to set into motion. It is developed for the child's body awareness, to experience that he or she can make the cradle move.

The free way in which the cradle moves helps to stimulate the child's sense of orientation. The sense is in the inner ear and it is not only crucial to our balance. It tells us about the body's position in relation to the surroundings and about direction and speed so, for example, we know how we should place ourselves when we are to catch a ball. It also helps the brain to sort out the many impressions so we can concentrate better.

The sense of orientation also contributes to strengthening the body's muscles in relation to balance. If the sense is poorly developed, the child has difficulty sitting on a chair and it can be difficult to concentrate on learning.

Leander Junior Bed

Leander
W70mm x L150 cm

对于父母而言，最幸福的事莫过于陪着自己的孩子长大成人。

Leander Junior 儿童床长 150 厘米，独特的 Leander 圆形设计为孩子带来无尽的想象乐趣。

床体高度适中，保证孩子可以轻松上下；科学的空间设计为孩子提供更宽敞的游戏场地。

Leander Junior 儿童床可以与新生儿婴儿床同时使用。

To see your child grow and develop into a little personality is a privilege for all parents.

The Junior bed is 150 cm long and it has Leander's round shape and design which stimulates your child's imagination and inspires play.

The bed is not too high so the child can easily get in and out of it. The bed does not take up much space in junior's bedroom so there is more floor space for games and activities.

The bed is ideal when the baby bed is to be made ready again for a little brother or sister.

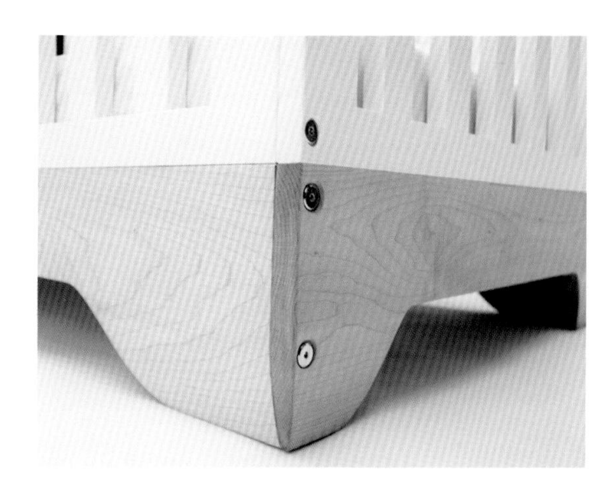

Echo Crib

KALON STUDIOS

经典造型，豪放与婉约兼修的气质与周围环境完美融合。

纯天然材料为孩子打造清洁的自由空间，床体表面采用百分百枫木精雕，做工细致，形神一体。

本款婴儿床力求简约、实用、美观，迎合现代人的审美理念。

森林管理委员会（FSC）认证、枫木制造、传统工艺定制，销往北美的产品由新英格兰制造，销往欧洲的产品由欧洲本地制造。环保、无毒木油安全无污染。

Celebrating the quintessential form of the crib. Balancing bold and delicate, the piece stands out while blending seamlessly into its environment.

One of the most important considerations of the design is the use of natural material and the creation of a clean, open space for the child. Each surface is gently sculpted from 100% solid maple, lending a subtle, unified quality to the form.

Driving forces of the design are simplicity, practicality and beauty with the intent being to find the intersection of these qualities. It is in this place that we simultaneously discover and embrace our perceptions of the contemporary.

Made of 100% FSC Certified Maple and made to order by traditional craftsmen. Pieces for North America are made in New England. Pieces for Europe are made in Europe. The wood is finished with a kind of non-toxic wood oil developed to be safe for humans and the environment.

Echo Toddler Bed

KALON STUDIOS
L131cm × W69cm × H10-15cm

独特简约的构造、精雕细琢的圆滑平面使顶部与底部呈现出自然的倾斜角度，摇篮的视觉感使整体造型更加整齐亮丽。

Echo Toddler Bed 为孩子精心打造自由空间，床高适中，孩子们可以轻松自信地爬上爬下，床头床尾营造的空间意象如同孕育胎儿的子宫般温馨，令“摇篮到童床”的经历更加美好。

FSC 认证，百分百枫木材质，新英格兰传统工艺，美国定做，无毒木油安全环保。

A singularly simple piece. Each surface is gently sculpted so that the piece slopes in at both the top and bottom, giving a soft cradling effect and lending a beautiful, unified quality to the form. Edges are rounded over and soft.

One of the most important considerations of the design is the freedom the bed gives the child who uses it. Graduating from a crib to a bed is a milestone for children that should be fully experienced. For this reason, the bed is set low to the ground allowing them to climb in and out with ease, building confidence and a sense of independence. The sides also clearly define the space of the bed, in a womb like way, as the child's own.

Made of 100% FSC Certified Maple and made to order in the USA by traditional craftsmen in New England. The wood is finished with a non-toxic wood oil we developed that is safe for humans and the environment.

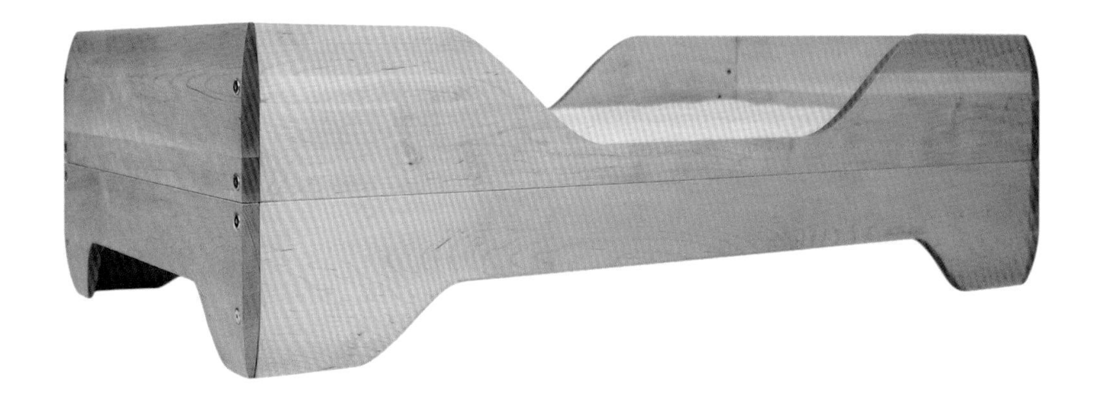

IoLine Bed

KALON STUDIOS
Toddler: Basic L137cm × W75cm × H20cm
Squared/Sloped L137cm × W75cm × H48cm
Twin: Basic L194cm × W104cm × H29cm
Sloped/Squared L194cm × W104cm × H67cm

一款为宝宝量身定做的幼儿床，高度适宜，造型整洁现代，方便孩子轻松上下，更添自信与独立。平台床两头搭配装饰性床头板，确保顽童安全。床侧有两种设计风格：直角简洁单纯，坡面精细复杂。简约造型与童真图案不但能提高安全系数，更为孩子的想象空间注入活力：置身大船之上，瞭望远方，隐身于丛林藤蔓之中……

可作为过渡床、幼儿床或双胞胎床。

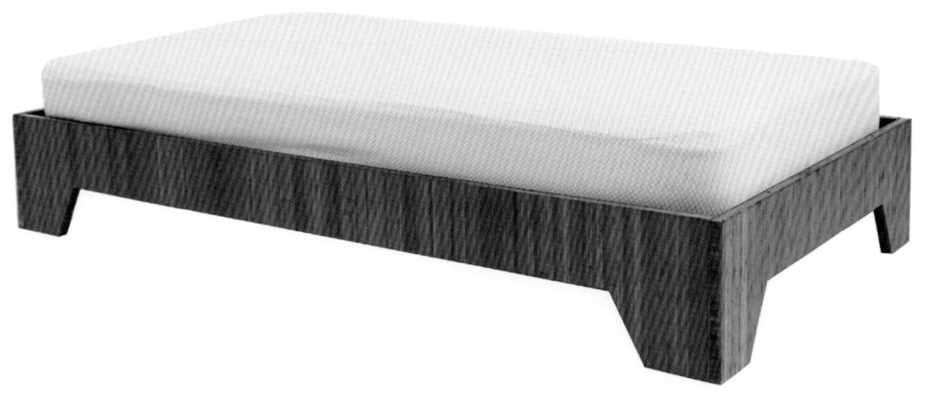

A bed that's your toddler's size. Set low to the ground, the clean, modern shape allows growing children to climb in and out of bed with ease, building confidence and independence. The IoLine Toddler Bed is a modern, platform bed but is also available with decorative sides and a headboard for wigglers. Sides come in two styles, squared for the purist, sloped for the sophisticate. The simple shapes and childlike cutouts of the sides provide more than just safety, they create a playful and engaging space for a child's imagination. A ship with portals to peek through, jungle vines to hide behind...

Available as a conversion kit, toddler bed or twin bed.

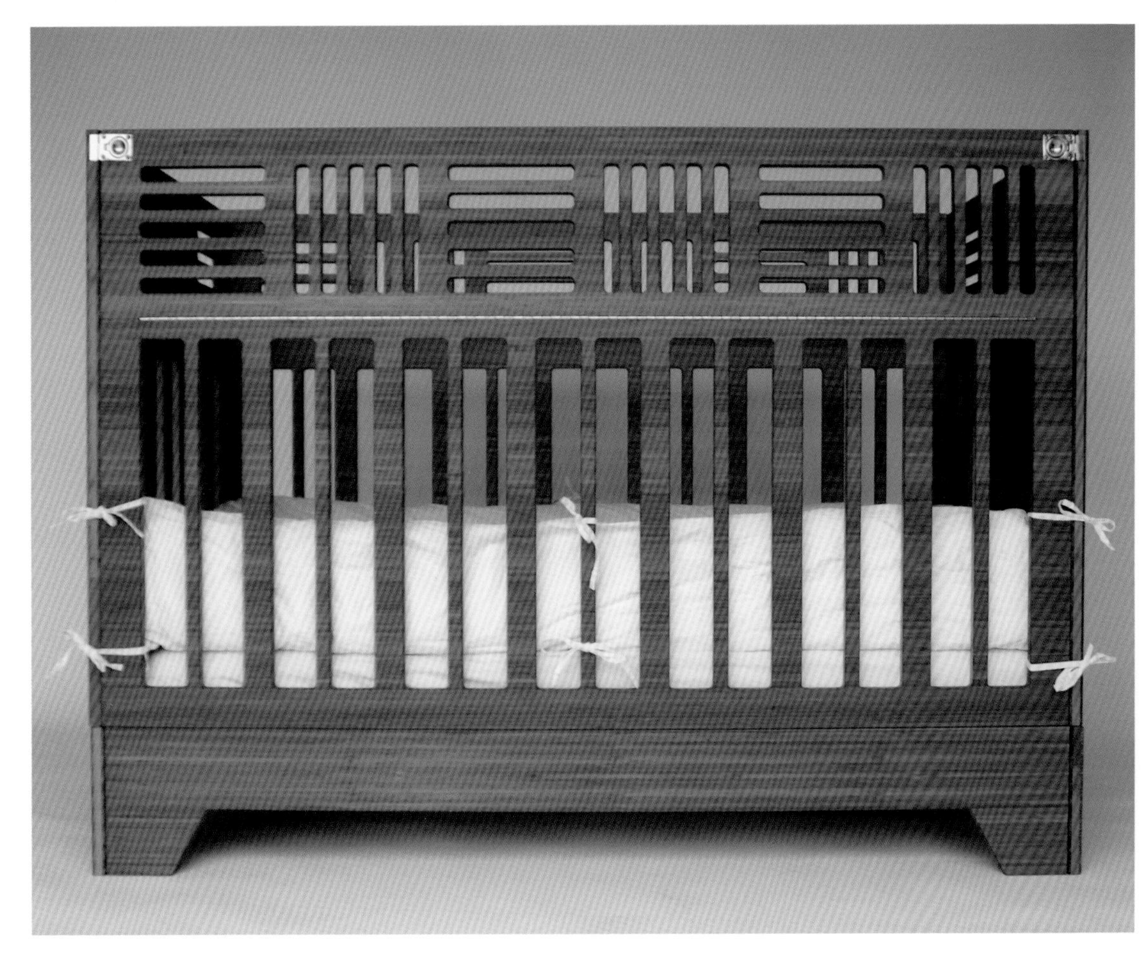

IoLine Crib

KALON STUDIOS
D75cm × W137cm × H100cm
D75cm × W137cm × H86cm

Io 婴儿床致力于伴随孩子的身心共同成长。四面可开放式设计保证童床内外视域开阔，如需改装成幼儿床，无需额外配件，可增加 6 年使用期限。

传统的木质屏风尽显灵性，造型优美的床架为宝贝儿打开一扇新世界的窗户。为孩子量身打造的 Io 婴儿床立足锻造他们的无尽想象力、创造力和自由奔放的表现力。

世界级顶尖绿色环保品质，传统工艺，打造可持续竹产品，销往北美的产品由新英格兰制造，销往欧洲的产品在欧洲本地制造。环保、无毒木油安全无污染。

Designed to grow with your child. Body, mind and soul. Open on all four sides the crib provides 100% visibility from inside and out. When it's time, the crib easily transforms into a modern, low-lying toddler bed, extending the life of this piece to 6+ years. No additional kit is required.

Traditional wooden screens were an inspiration for this piece. The design seeks to create a beautiful and engaging frame through which the newborn and developing baby sees the world. Originally made for the designers' own child, the Io crib seeks to capture a child's unbridled and pure imagination, creativity, performance and wild spirit.

Consistently recognized as one of the top green products in the world. Sustainably made from bamboo by traditional craftsmen. Pieces for North America are made in New England. Pieces for Europe are made in Europe. The wood is finished with a kind of non-toxic wood oil developed to be safe for humans and the environment.

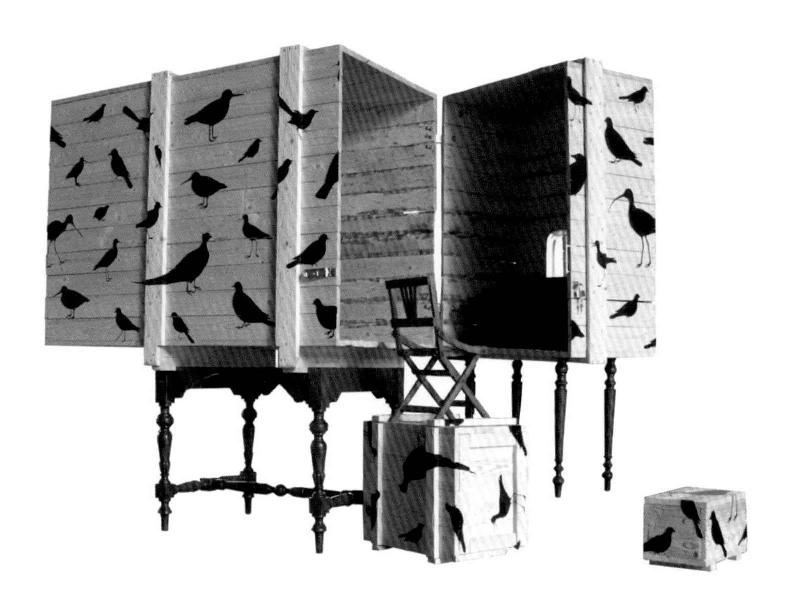

Birdwatch Cabinet Boy

Studio Makkink & Bey BV
Photographer: Studio Makkink & Bey

“男生观鸟阁”是一个七岁男孩的“卧室”，由一张旧桌、一张办公椅及运输箱组合而成，喷砂处理令运输箱光洁如新。

园亭、浴炉表和观鸟阁属旧家具收藏系列产品，从中人们不断开掘出新产品和新功能。

The birdwatch cabinet is a little sleep maisonnette for seven-year-old boy. An old table and office-desk form together with their wooden traveling box a research-worker-cabinet. Through sandblasting the traveling box gets a rich finishing.

Gardenhouse, bathstovetable and birdwatch cabinet, are a series of products in which collected old pieces of furniture are combined, in search of new products and functions.

Birdwatch Cabinet Girl

Studio Makkink & Bey BV
Photographer: Studio Makkink & Bey

“女生观鸟阁”是一个七岁女孩的“卧室”，由一张旧桌、一张办公椅及运输箱组合而成，喷砂处理令运输箱光洁如新。

园亭、浴炉表和观鸟阁属旧家具收藏系列产品，从中人们不断开掘出新产品和新功能。

The birdwatch cabinet is a little sleep maisonnette for seven-year-old child. An old table and office-desk form together with their wooden traveling box a research-worker-cabinet. Through sandblasting the traveling box gets a rich finishing.

Gardenhouse, bathstovetable and birdwatch cabinet, are a series of products in which collected old pieces of furniture are combined, in search of new products and functions.

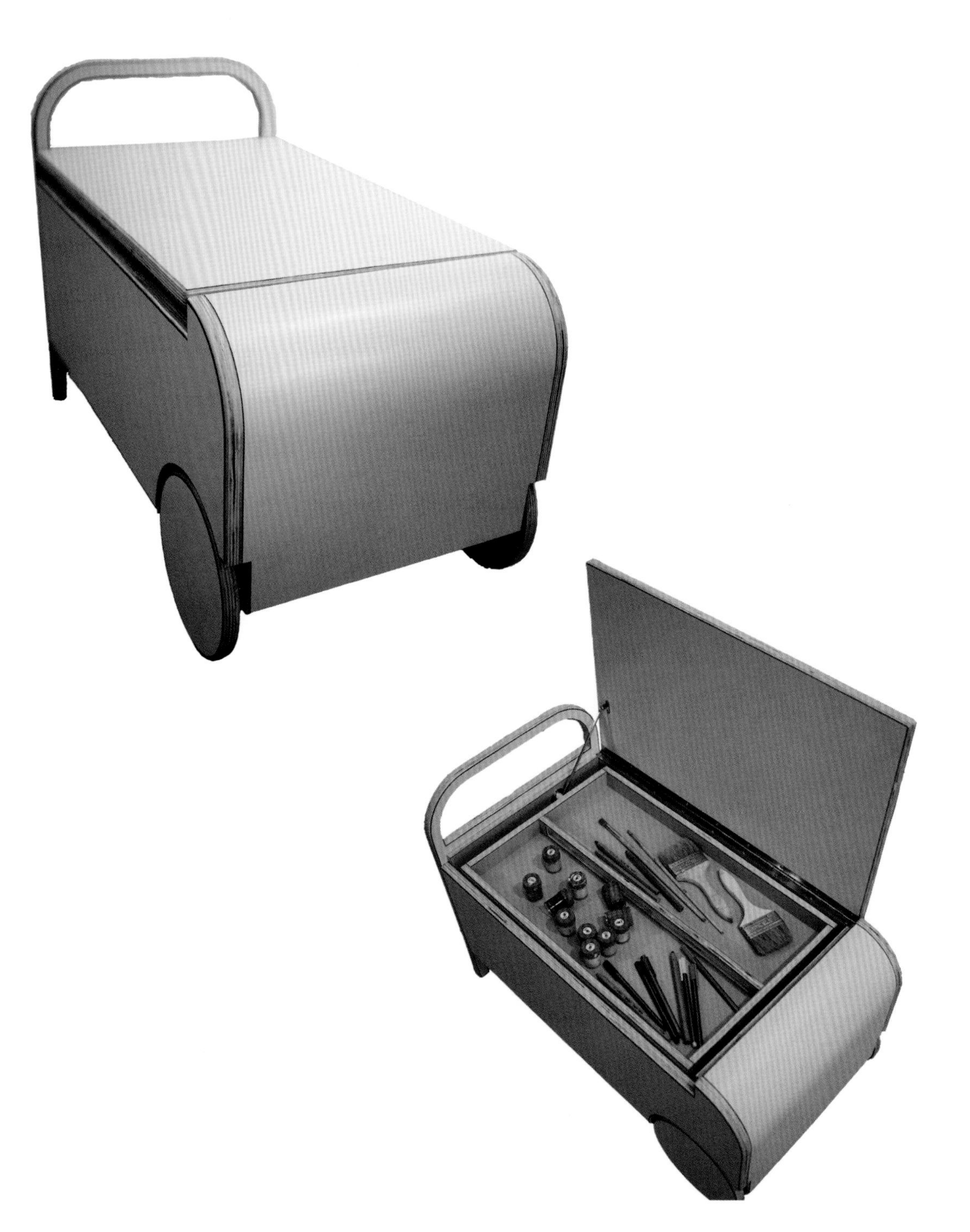

IO Arts & Crafts Doodle Box

IO KIDSDESIGN
L75cm × W52cm × H40cm

手提存储箱专为满足孩子学习、绘画和创造兴趣而设计。箱底滚轮便于自由移动，箱体可做长凳使用，内部设有艺术作品存储空间和放置绘画用具的可移动托盘。箱内纸卷拉伸后与箱盖形成自然画板。

A portable storage unit designed to accommodate children's interest in learning, drawing and creating. The unit is designed with integrated wheels that allow children to easily move it around. It also doubles as a bench seat with internal storage for art work and removable tray for drawing and painting accessories. The box accommodates a roll of paper which when pulled over the lid creates a drawing board.

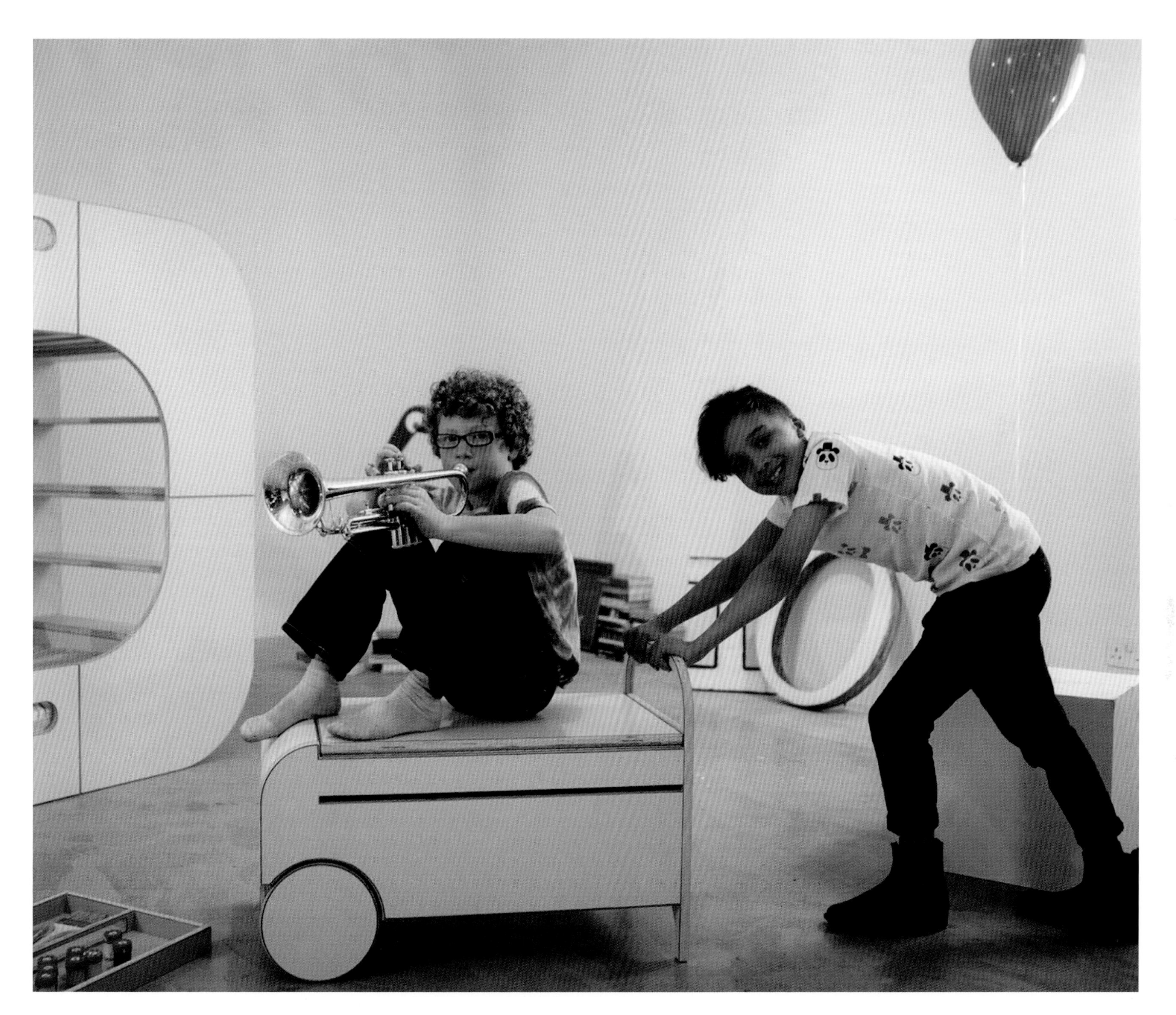

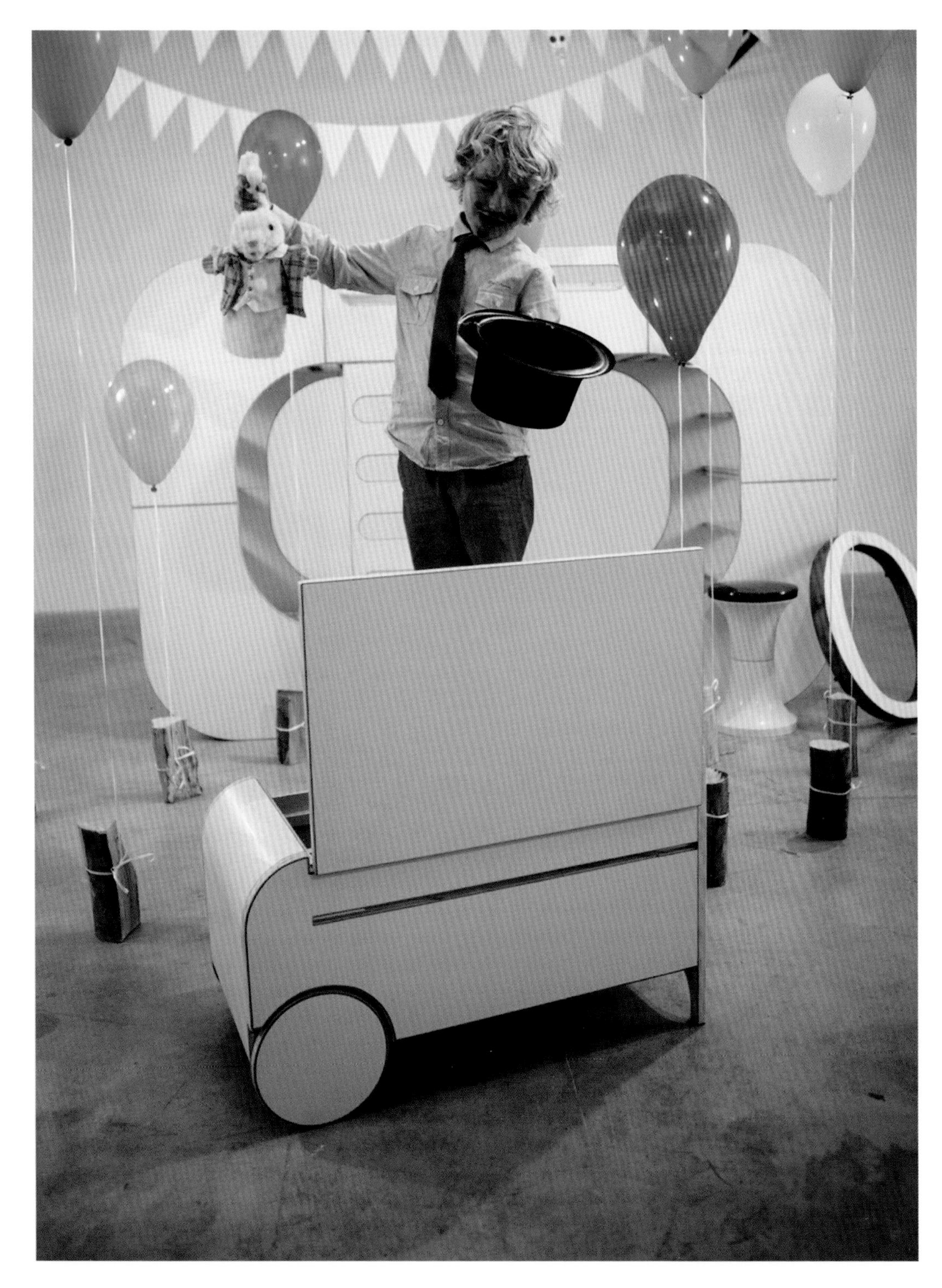

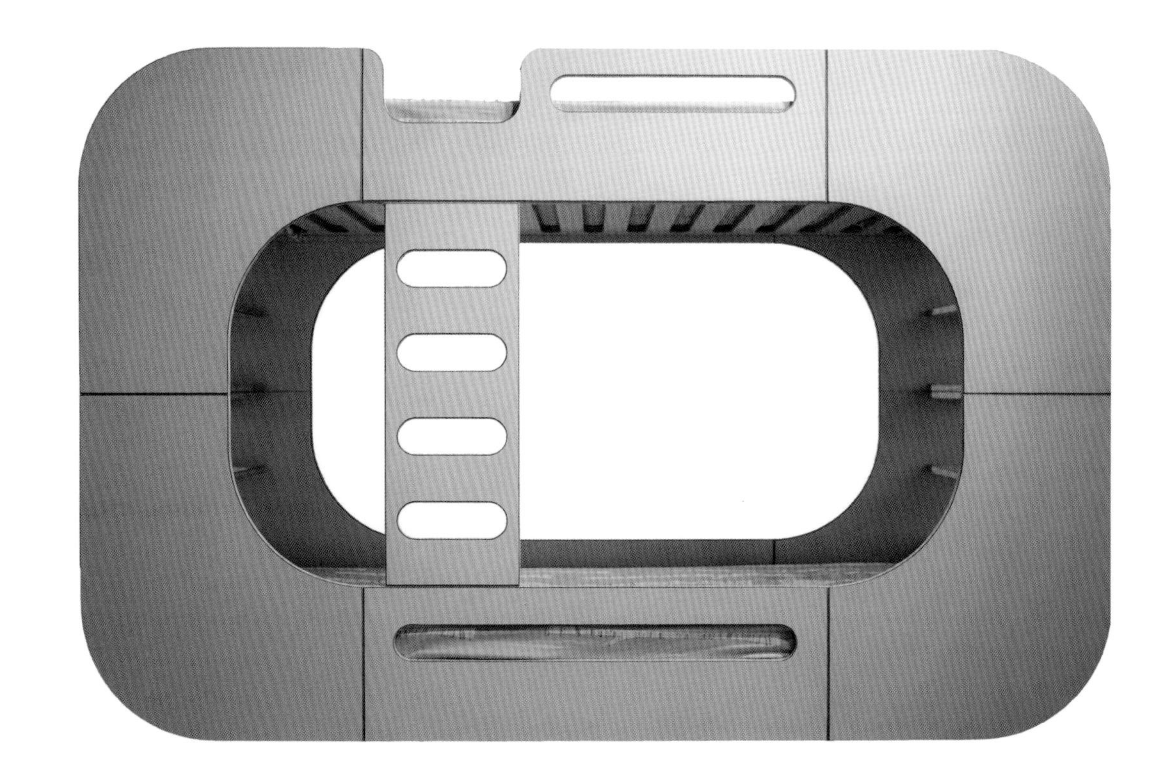

IO Bunk Pod

IO KIDSDESIGN
Recommended mattress size: 90cmx190cm
Recommended height of mattress: 10cm
Dimensions: W94cmxL230cmxH150cm

“多效合一”的整体设计理念为孩子节约更多睡眠、存储和学习空间。Bunk Pod 立足打造全能高端产品，一直严格遵守英国与欧盟的安全标准。床体构造简单，易于组装改造，由两个独立单元、沙发床和存储工作台组成。

An all-in-one space saving place for children to sleep, store and study! Our versatile, high-end Bunk Pod has been designed to comply with UK and European safety standards. It has a minimum number of components and can be easily assembled and transformed to perform a selection of functions with minimum effort including two stand alone units, day bed and work station with storage.

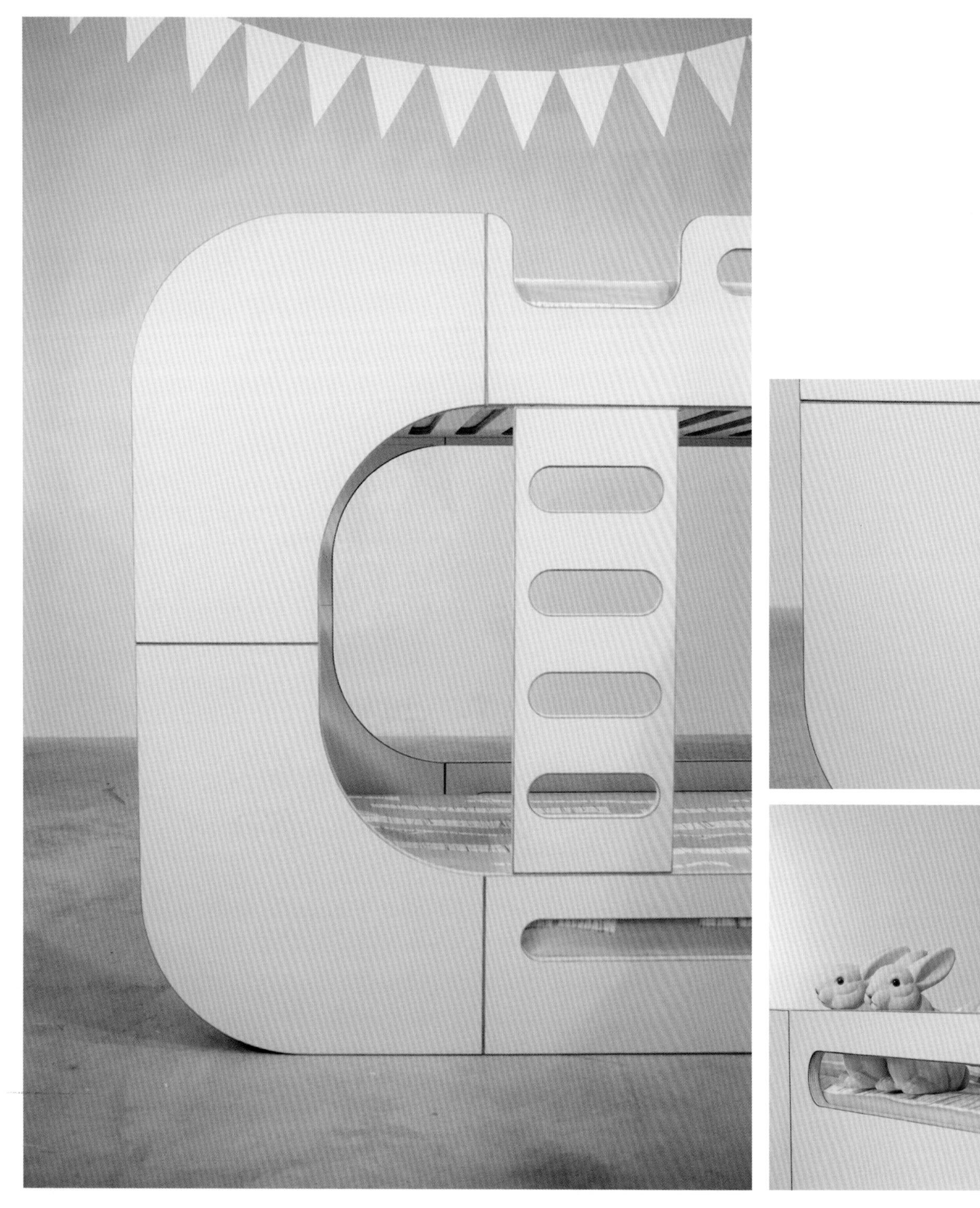

吃 的 设 计

Products for Eating

除了睡眠之外，儿童在清醒时的一切用品都可以作为教具来对待，这是刺激神经发育、锻炼感觉、训练控制能力、身体生长和学习人类生活方式的重要媒介。

就餐具而言，不同年龄的儿童对其理解是不同的，儿童对奶瓶、水杯、餐盘、餐碗、手、餐勺、筷子、餐桌凳的使用和理解是递进的过程，这些用具看似餐具，其实都是作为训练儿童成长的教具。婴幼儿时期，他们用手抓饭都很难准确地送进嘴里，对餐勺起初是不知道正反面和形状功能的作用，只是单纯模仿大人持勺的动作，对他们来说勺子和筷子是一回事，儿童就是这样不断地使用、练习才对勺面、勺背、距离有了认识和控制。

衣食住行是人类生活的基本内容，让儿童尽早介入全过程，培养他们社会角色意识很有好处。尤其对改变中国父母包办过多、剥夺孩子锻炼的机会、代替孩子成长等溺爱现象是非常有益的。只要大人们能够控制住局面，看似较危险的情况不妨让孩子试一试，分清有大人看护与儿童独自玩耍的安全分寸，孩子受点轻伤和疼痛都会有益于他们健康成熟。大人对儿童成长的基本态度就是不能代替他成长，不能让他在襁褓中长大。

本章列举了儿童餐具的设计，亦是一种对生活方式的体验。

When your babies are awake, you may turn everything available into teaching tools helping to stimulate neurogenesis, sharp the sense and develop the discipline of the children, more importantly, to pave the way for their future life.

Children of different ages may acquire new understanding toward the same tablewares like nursing bottle, cup, plate, bowl, hands, spoon, chopsticks, dining table and stool, so in most cases, these dinnerwares could be applied as teaching tools to train children. For example, an infant is usually complete ignorant of the shape or function of a spoon when he or she even cannot eat with hands, so they can only imitate the adults while failing to differentiate between spoon and chopsticks. But little by little, children learn to understand and use spoon through repeated trying and training.

As clothes, food, shelter and travel are basic elements for life, it is important for children to be actively engaged in family life as early as possible so as to learn to shoulder social responsibilities. Also the participation is specially beneficial to Chinese kids whose parents tend to take care of everything and even make choices for them. In truth, parents are encouraged to let kids go and give them free space to take risk but keep it in control and in this way, minor injuries and pains may turn out to be a blessing in kids' development. After all, they cannot be tied to mother's apron forever.

To introduce you a different life style, this chapter lists some designs of kitchen utensils for children.

FunFam

funfam

采用具有很强的抗菌性，轻而坚固，对环境无害的天然素材竹子的 FunFam 餐具，考虑到了提高食欲方面的设计，发挥了竹子特性的餐具非常轻巧，从初食儿童到握力减弱的高龄者都可以使用。日本的工匠，一件一件精心削做的 FunFam 竹餐具增添了餐桌上的温馨和色彩。

With the use of bamboo, a light but strong natural material, FunFam is well received for its antimicrobial properties, harmlessness to environment and appealing to appetite. The light utensils are the ideal choice of infants and the elder with decreased gripping power. Handmade work by Japanese craftsmen adds color and warmth to your dinner table.

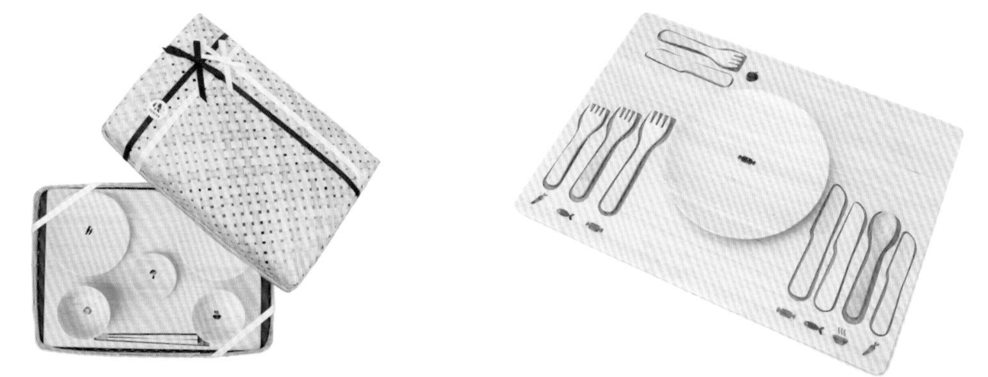

寿

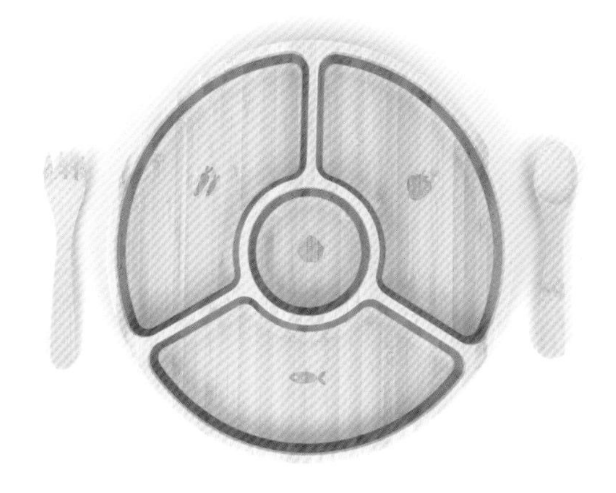

LIVING PRODUCTS CONCEPTS STYLE , TEXTILE & SURFACES

Charlotte Skak

研究表明，那些参与家庭餐准备过程的孩子们长大后会拥有更加健康的饮食习惯，对未知食物也会怀有更强烈的好奇心，而这些习惯往往会延续至成年，因此，与孩子在厨房共度时光对他们的健康成长大有裨益。

“厨房孩童”厨具系列包括一把刀、一把蔬菜削皮刀、一个蔬菜刨丝器、一个碗、一个用于浇汁的混合瓶、一块案板以及一副保护手套（家长可放心让孩子使用锋利物品）。

锋利的工具采用不同颜色和样式，从而区分安全抓握位置及危险区域。所有把手由橡胶制成，并有特色防滑图案以确保安全抓握，这些功能图案看似整体构造的一部分，实则产品的条形码，即每款产品的电子标识。

Children engaged in the preparation of the wholesome family meal acquire healthier eating habits and a greater curiosity towards new and unknown foods because they take the ownership of their meal. These habits continue into adulthood, leaving the extra time spent in the kitchen with your child as a very good investment in the health and future life of your child.

The Kitchen Kids Collection consists of the following: a knife, a vegetable peeler, a chopping board, a vegetable grater, a bowl, a bottle for mixed condiments, and a protective mitten, giving the parents peace of mind when the children use the sharp tools.

Using color and form, the sharp tools are designed so there is a clear distinction between where it is “dangerous” and where it is safe. All handles have a distinctive functional pattern and a rubberized surface providing the best grip possible. The functional pattern is at first sight just a structure, but the structure is derived from a barcode, the digital signature of the specific product.

游 戏 的 设 计

Products for Gaming

儿童游戏类用具很难界定，儿童学习都是在游戏的状态中进行的，本书将书桌椅、玩具均归于此章正是出于这种思想。

玩具是儿童游戏的媒介，儿童可以将任何东西变成玩具，他可以把一个小凳翻过来当成车，把很多凳子连起来当成火车，把枕头当成娃娃，把饭桌当成房子……孩子的异构能力是超常的，他们会轻而易举地将完全不同的东西建立联系，儿童的这种举动正是追求整体认识事物的表现。认识事物之间的关系才算认识事物，才算懂道理。儿童从不只关心知识点，他们往往从某个问题（知识点）出发关注的是“为什么”，“为什么”就是事情的原委、来龙去脉、要素之间的关系。当你为孩子倒杯水喝的时候，他会想到：是热的吗，甜的吗，是我喜欢的小花猫杯子吗，还会洒在床上吗？每个问题都与他经验相关又会启发下一个疑问。疑问是儿童的生活内容，疑问正是儿童激情活力的表现，儿童的生活中充满了疑问，儿童从来没有所谓专业的界限，从不管有用没用，只要是问题就不排斥，都要去追问。正是这些无数的疑问让儿童不知不觉地不费吹灰之力神速进步。

儿童玩具的类型要充分考虑其年龄成长特点，即功能特点。如，不论训练动作、视觉、听觉、注意力、认知等都不能仅从知识点出发，而要从成长发育和培养良好习惯出发。

启智类玩具的设计关键要制造疑问（悬念），并不是提前学会小学课本。总之，游戏类用具要关注儿童身心健康成熟和发展能力，让孩子懂得什么时候做“老大”，什么时候当“老二”，什么时候专注，什么时候随意，什么是合作，什么是交流等。

It's always hard to define what "the game appliance for children" is, because most of the time children acquire knowledge during playing. According to this, desks, chairs and toys are included in the same chapter.

Toys are the media of children's games. Children can turn nearly anything into a toy if they want. They can make a car by putting a bench upside-down; they can link several benches together to make a train; they can turn a pillow into a doll and turn a table into a house... Children's creativity is always beyond imagination and they can easily connect two seemingly different things together, which shows their ability to view the world as a whole. Recognizing the connection between things is what we count for truly recognizing the world or what we call "acknowledgeable". Rather than concerning the individual point, they are more curious about the underlying mechanism of a question (knowledge point) and ask "Why it's going this way?". Such "why" questions stand for the reasons or mechanics behind the phenomenon and the links of different elements. When you bring your child a cup of water, he will wonder "Is it hot or sweet?", "Is it served in my favorite kitty cup?" or "Will I spread it on the bed?" Every question like this related to their personal experiences will stir the next question, which is the essence of children's life, life with curiosity, in another word, passion and creativity. Questions from children may cover various fields whether they are useful or not. And it is such an endless curiosity that brings them into maturity in an easy and progressive way.

If you are choosing toys for your children, you do need to take their age into consideration and make your decision based on the function of toys. For example, all training of behavior, vision, audio, attention, cognition and etc., is not only built from the separate knowledge point, but also from the natural process of growing and cultivation of good habits.

The design of intelligent toys focuses on generating curiosity rather than teaching knowledge in the primary school to children in advance. All in all, the game appliance must pay attention to children's physical and mental health as well as developing their skills. It should teach them and make them understand the behavioral rules like when to be a leader and when to be a follower, when to be concentrated and when to be at ease, what cooperation is and what communication means, etc.

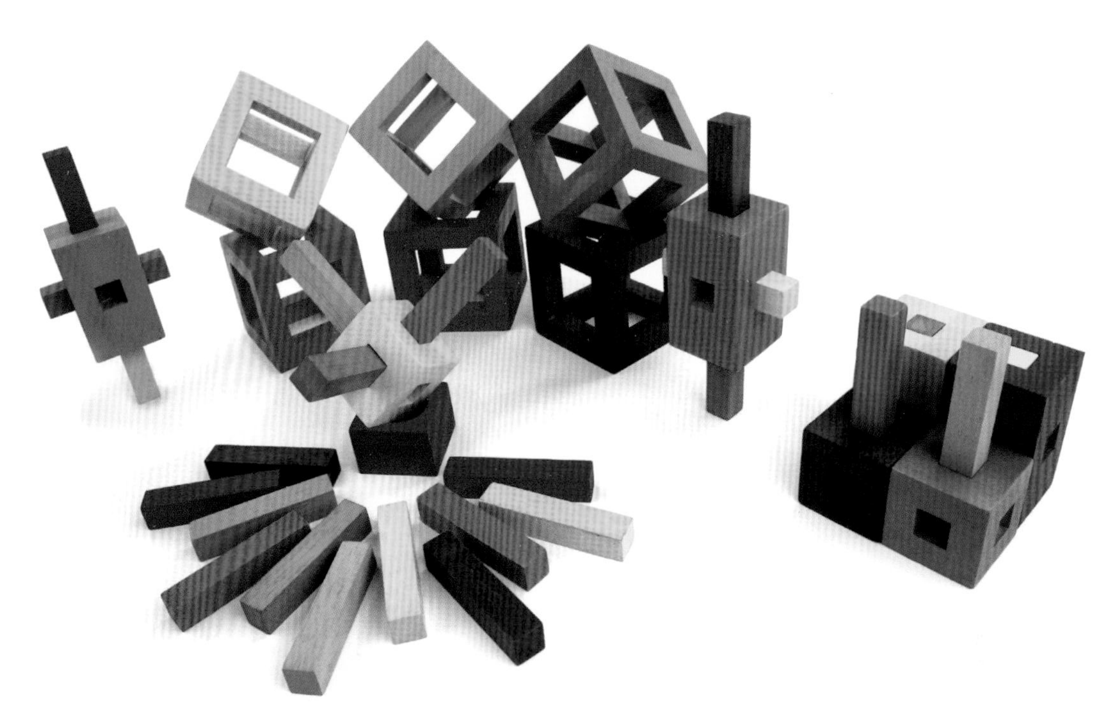

Twig

Fat Brain Toy

Twig 积木有着精美的木材形状和充满生气的色彩。天然木材经过精心雕刻和用心上色，橡胶木上涂水溶性涂料。光线通过 Twig 的窗户产生独特效果——线条明亮，轮廓清晰，结构合理。

Twig 积木倡导在自然的游戏状态中学习，锻炼儿童精细动作能力的同时，也能帮助提高孩子的各种能力，如逻辑分析、解决问题、空间能力、创造和发明力、感官意识、言语和词汇流畅性、协作和独立玩耍，还有艺术性的表达能力等。孩子们喜欢将这些独特的小木块组合成各种形态……而成人则会爱上 Twig 带来的创作自由。无论在游戏室、办公室还是咖啡桌上，它总是一样地吸引人。Twig 设计属于每个人！

Precision wooden shapes and vibrant colors meet in modern building blocks. Natural wood is carved with care and colored with brilliance. It is a rubberwood and water based paint. Light plays its way through the windows of Twig — illuminating lines, contours, space, and arrangement.

Learning comes naturally. Twig construction sets refine small motor abilities. Twig helps develop skills in logic, problem solving, visual-spatial ability, creativity and ingenuity, sensory awareness, language & vocabulary proficiencies, cooperative and independent play skills, and artistic expression. Kids love putting the unique interchanging pieces together.... adults love the freedom of design. Equally appealing in the playroom, in your office, or on the coffee table. Twig is design for everyone.

Dado Cubes

Fat Brain Toy

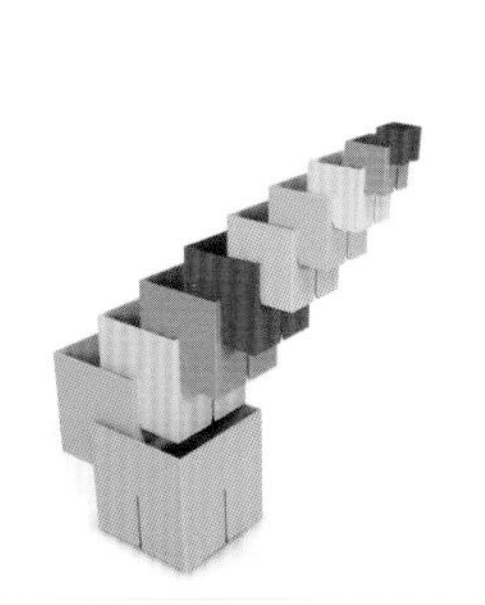

当你试图探索它的建筑原理时会发现，无论在比例、平衡、结构，还是色彩方面，Dado 都堪称艺术与科学的完美结合。这是一个传统建筑积木类玩具领域的全新转折。Dado 每一个小立方体都在充分调动你的想象力，这些环环相扣的小积木让你能够创造无限数量的三维结构。既可以水平搭建，也可以垂直建造，甚至镶嵌在一起。Dado 让你在设计精巧、抓人眼球的游戏享受中拓展视觉空间，提高解决问题的能力。

Dado combines art and science as you explore architectural principles... proportion, balance, structure and color. A new twist on classic building blocks, Dado engages your imagination as the slits on each cube are interlocked to create an unlimited number of three-dimensional structures. Constructed horizontally, vertically or nested together, Dado invites visual spatial development and problem solving through design-centered, attention-grabbing fun!

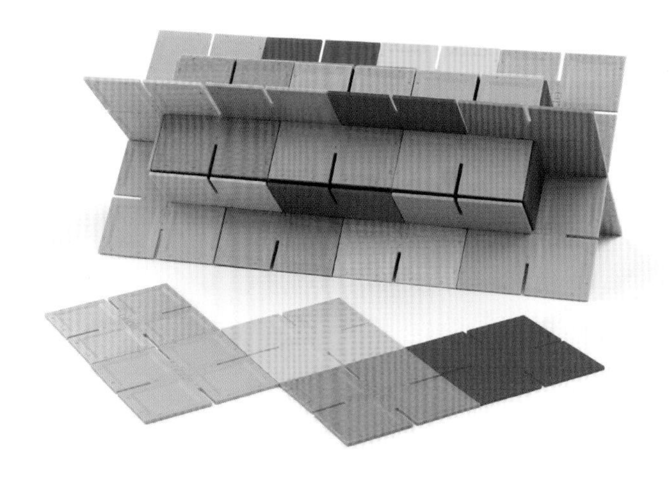

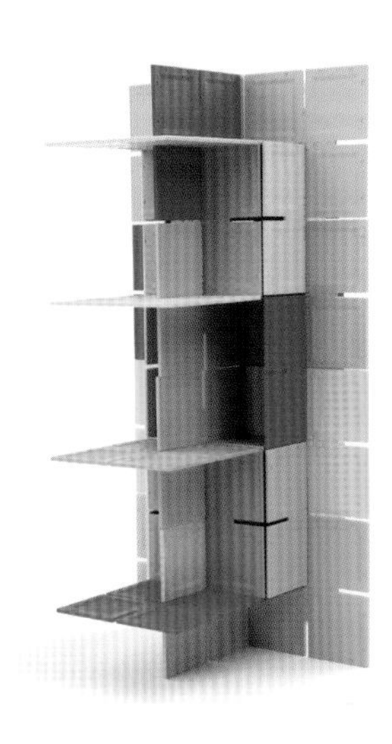

Dado Squares

Fat Brain Toy

Dado Squares 是 Dado 玩具战线的又一新成员，将复杂的思维理念通过简单的设计表现出来。不论是否喜欢艺术和科学，孩子都能在探索建筑原理的过程中体验比例、平衡、结构和色彩的奥妙。Dado Squares 一定能激发孩子的灵感，满足孩子的娱乐需要，并在利用这个简单的玩具创造出任何三维结构的同时拓展视觉空间能力和解决问题能力。

Dado Squares are the newest addition to the Dado toy line inspiring complex thought through its simple design. Whether you favor art or science as you explore architectural principles ...proportion, balance, structure and color, Dado Squares are sure to stimulate and entertain your mind. Create any number of three-dimensional structures and explore visual spatial development and problem solving with this simple toy.

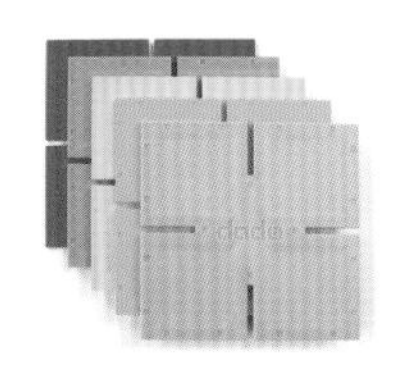

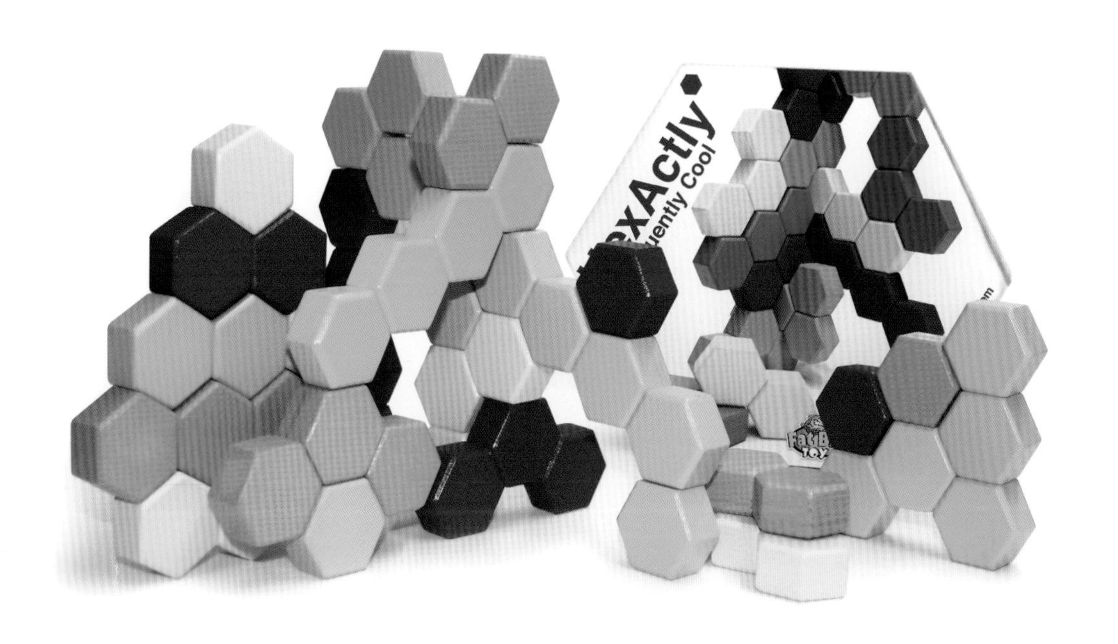

HexActly

Fat Brain Toy

五种独特的六边形基础部件共同构成一个多功能、多式样，并能拓展思维的建筑积木。这五种六边形的基础部件包括一个单一的六边形，两个、三个和四个六边形的线性组合，三个六边形的非线性组合。

这款建筑积木产品所带来的思维训练比它的结构看起来更加不可思议，而 HexActly 所能激发的缜密思维、创造力以及视觉空间活动同样意义非凡。快试试你的组装技巧，成为 HexActly 现代建筑砖块积木的初级玩家、中级玩家，直至成为“Hex”的结构专家。

One of five unique hexagon-based pieces make up a single block piece providing versatility, variety, and extended thought. The five hexagon-based pieces include: a single hexagon; 2, 3, and 4 hexagons - joined in linear arrangement; 3 hexagons joined in a non-linear arrangement

The rewards of this block play go beyond incredible-looking structures. The complex thought, creativity, and visual spatial activity HexActly blocks inspire is worth noting. Try your skills in assembling HexActly modern building blocks into beginner, intermediate, and "Hex"pert structures!

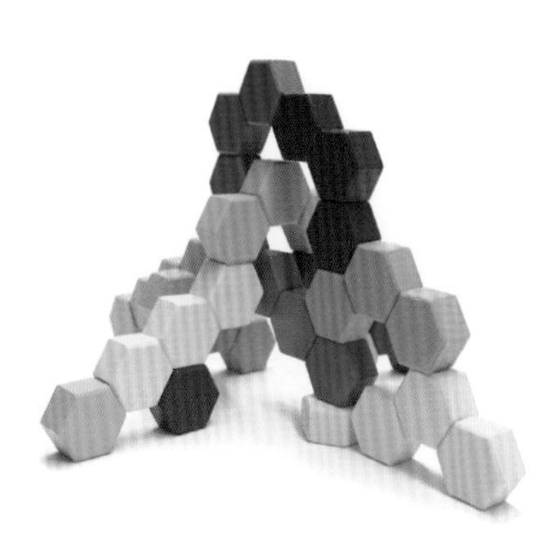

Reptangles

Fat Brain Toy

这套高度优化的建筑玩具是锻炼想象力的完美选择。爬虫式的有趣造型会使孩子们很快爱上这款建筑玩具套装。入门阶段易如反掌，自然过渡到更高水平的益智训练，逐级调动你的想象力和创造力，寓教于乐。一旦开始游戏，孩子们很快就会忘记他们在玩的是经过精心设计来提升智力的玩具。

Rectangles 的核心部件是多面几何体。这种形状使得 Rectangles 的所有小部件都能以无限种方式组合起来。通过从简单到复杂构建各种形状的多面体，孩子们的建筑技巧会很快得到提升。这些几何组合不断激发着孩子们的设计灵感。

A highly evolved construction toy, Reptangles are a perfect fit for building intelligent imaginations! Because of their reptilian resemblance, children are immediately drawn to this building toy system. The snapping begins and soon leads to new levels of imagination, creativity and educational play! Once they start creating, children forget these are turtles with purpose!

The geometry of the polyhedra shape is an essential component to the appeal and application of Reptangles. This shape is key to the countless ways the turtles can be combined. Children quickly advance their skills by building simple to complex polyhedra, the geometrical configuration which inspired their design.

Tobbles

Fat Brain Toy

Tobble 由六个重量各异的球体结构组成，这些小球可相互镶嵌，能构造出多种富有巧思的多彩组合。

这个堆叠玩具蕴含的物理学设计原理可以帮助孩子认识体会平衡、比例、色彩和结构，同时促进他们的感官探索、手眼协调、视觉空间认知、平衡等各种能力发展。

Tobbles 会在孩子们的小手和聪明才智下演化出无限变体。一个精心设计的对话模块可以让孩子在游戏的同时享受探索和发现的乐趣。

Tobbles is six uniquely weighted spheres which nest inside each other and can be balanced in all sorts of clever and colorful combinations.

The interplay of physics and design results in a stacking toy that teaches balance, proportion, color and structure while encouraging sensory exploration, hand-eye coordination, visual-spatial awareness, balance and more.

Little hands will delight in the never-ending discovery of Tobbles. A design-savvy conversation piece which invites you to discover and explore while providing hours of abstract fun!

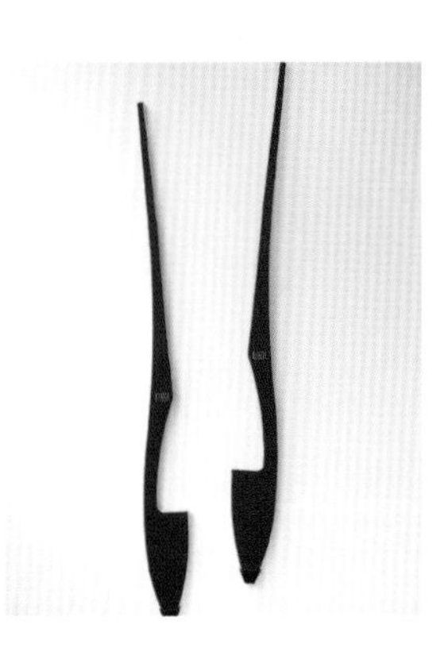

Toboggan

tau

Dimension as puzzle plate: 80cm x 75cm x 1.8cm

Dimension as assembled toboggan: 55cm x 42cm x 27cm

完美的 Toboggan 玩具专为一岁及以上并且喜欢坐雪橇的小朋友们设计。

The perfect Toboggan for little children ages one and older who love to let themselves be pulled through the snow.

Seesaw

tau
Dimension as puzzle plate: 150cm x 90cm x 1.8cm
Dimension as assembled seesaw-horizontal:
146cm x 60cm x 68cm

跷跷板将为家庭庭院添加新的乐趣，防滑地板保护设计同时适合客厅使用。

推荐两岁及以上儿童选用这款产品。

The seesaw is ideal fun for the garden. And with its non-slip floor-protectors, it's also suitable for your living room.

We recommend the seesaw for children ages two and older.

Wheelbarrow

tau

www.tau.de

Dimension as puzzle plate: 84cm x 75cm x 0.9cm
Dimension as assembled wheelbarrow:
53cm x 32cm x 40cm

Wheelbarrow 以寓教于乐的方式训练孩子的平衡和协调能力。孩子们也喜欢坐在里面，让自己围着家长转圈圈。

Wheelbarrow 适合两岁及以上儿童。

The wheelbarrow trains your child's balance and coordination in a playful way. Kids also love to sit inside and let themselves be wheeled around—by you!

The wheelbarrow is suitable for children ages two and older.

Black Beauties

INEKEHANS/ARNHEM

Black Beauties（黑美人）完全由黑色可循环环保塑料制成，具有防风、防水、防酸、防碱和防紫外线的功能。这种材料可以用于潮湿环境，如浴室台阶，也可用于秋千等户外环境。最初设计师们想把它设计成一款彩色产品，但后来发现大部分可用材料是白色、自然色或黑色。更重要的是，黑美人向我们证明了孩子们不仅对色彩有感觉，而且经常对物体的形状和游戏方式也有积极的反应。

All items are made of black recycled plastic: wind-, water-, salt-, acid-, UV-resistant. The material allows products like the bathroom step and the swings to go outdoor and into wet environments. The decision to make them in the black material came when designing colourful work in years before, and observing that most materials are available in white, natural and black. More importantly, the Black Beauties show that children do not only react to colours, but very often they respond to shapes, opportunities and ways of playing with things.

Parent Chair

MOOOI

为父母制作的儿童家具！谁说孩子的家具就不能带有家长的个人风格？Marcel Wanders 和 Moooi 已经为家长设计出了一件绝佳的居家伴侣，Parent Chair 不仅个性十足，且轻巧坚固。

Children's furniture for the parents! Who says children's furniture has to be mismatched from your own style, Marcel Wanders and Moooi has brought to the collection this great accompaniment for your home. The Parent Chair is light and robust. Perfect to play with and stylish to own.

Parent Table

MOOOI

为父母制作的儿童家具！ Marcel Wanders 和 Moooi 已经为家长设计出了一件绝佳的居家伴侣，Parent Table 不仅个性十足，且外观柔和，轻巧坚固。

Children's furniture for the parents! Marcel Wanders and Moooi has brought to the collection this great accompaniment for your home. The Parent Table is soft, light and extremely robust. Perfect to play with and stylish to own.

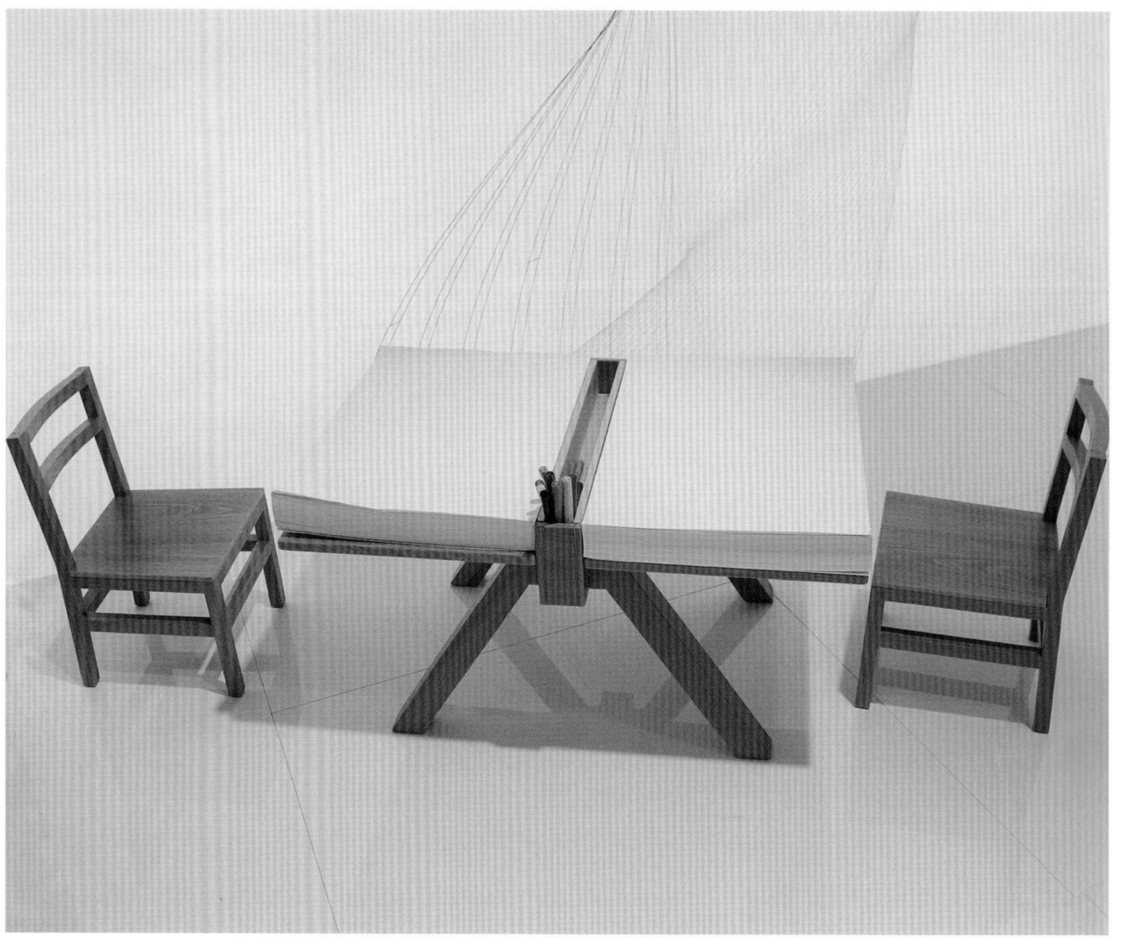

cuscino & foglio

domodinamica

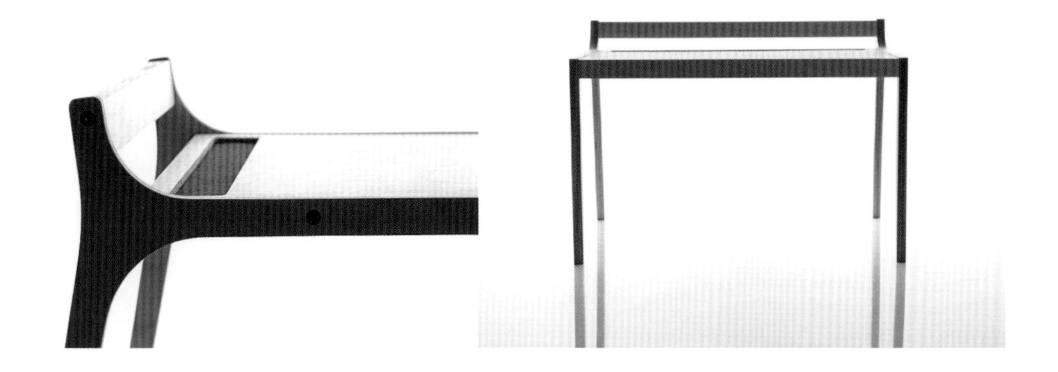

sibis afra

sirch

L57cm × H60cm × W87cm

sibis afra 由桦木制成，适合二至八岁儿童使用。

Plywood for children of 2 to 8 years of age.

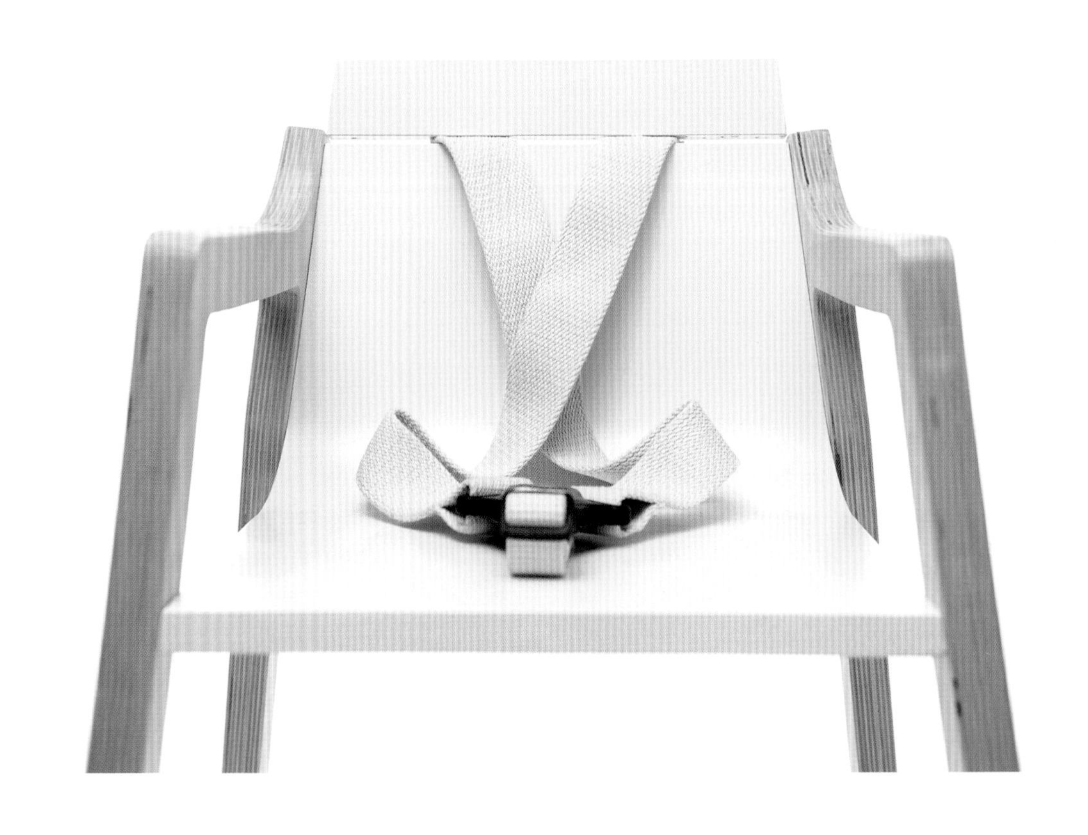

Beppo — Childrens High Chair

sirch

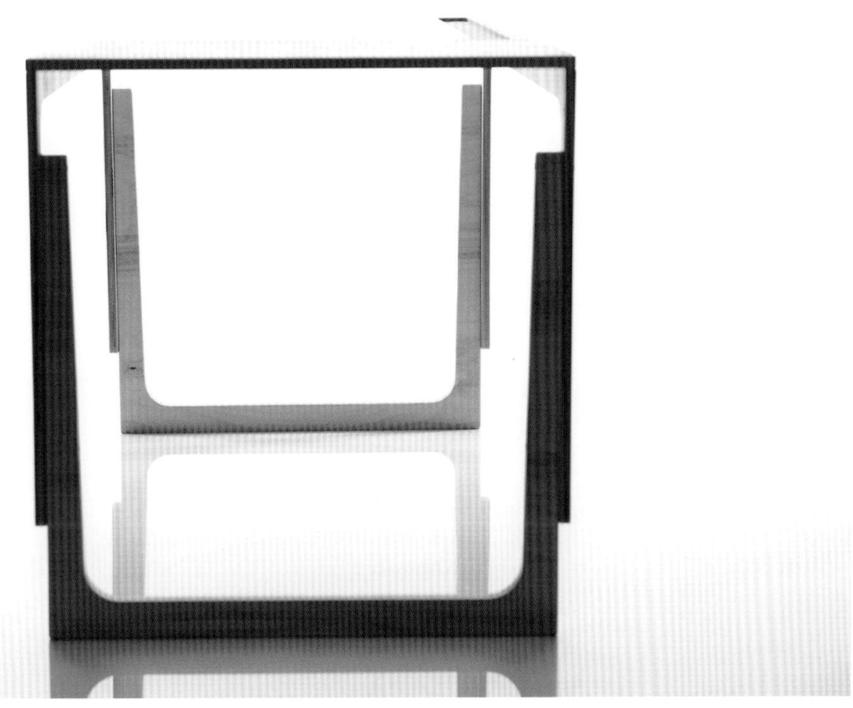

sibis Vaclav

sirch
Dimension: L115cmx W63cm
Height of writing surface: 60cm, 66cm, 72cm

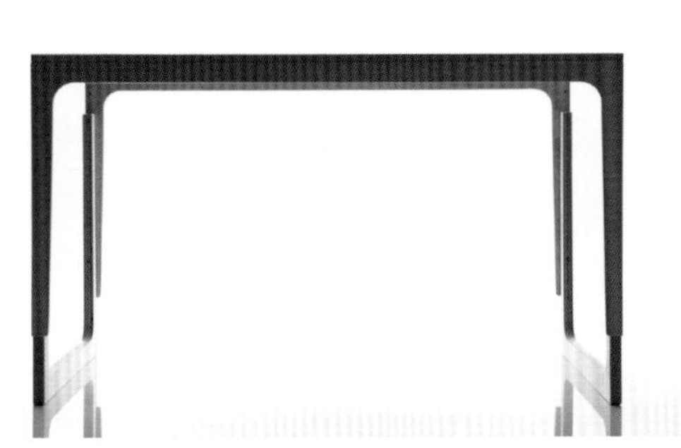

sibis Vaclav 由精细抛光的桦木胶合板原料制成，台面打蜡，并可拆卸。为六岁及以上儿童设计。高质量的铅笔槽搭配三种不同颜色设计。

Made of raw, finely-polished birch plywood, table top waxed white, the table can be dismantled. It's for children of 6 years upwards, pencil groove with high quality inlaying three different colours.

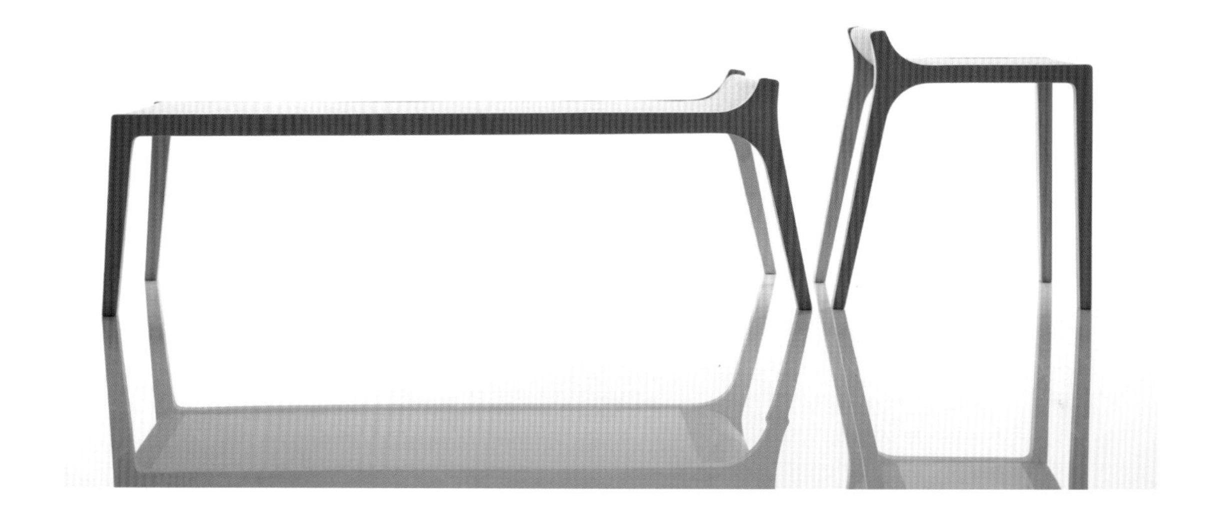

sibis erykah

sirch
Plywood for children of 2 to 8 years of age.
Dimension: L98cm x H31cm x B42cm
Height of seat: 27 cm

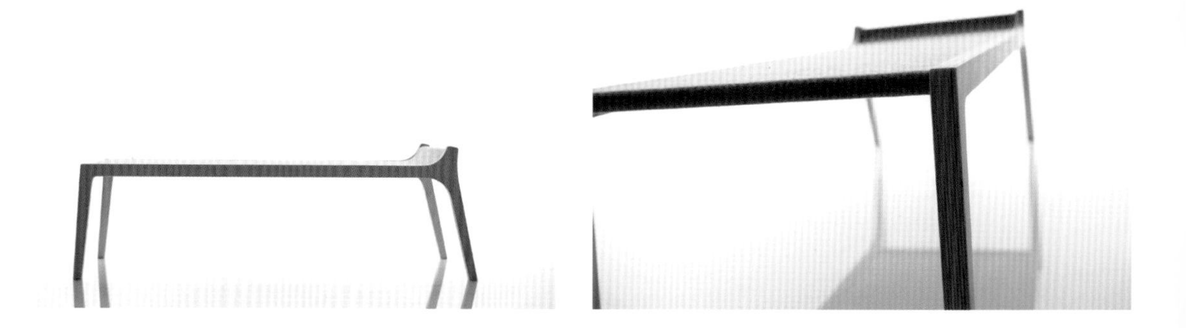

Tone Kids Chair

Leif. designpark
L28cm x W33cm x H48cm

Tone 儿童椅选择可持续木材制成，呈现自然颜色和独特纹理。坚实的美国黑核桃木框架支撑座椅，靠背采用樱桃木贴面的胶合板。设计结合了传统的日本拼花地板和现代的表现形式。

Tone Kids Chair uses a selection of sustainable timbers, flaunting the natural colour and grain differences. The solid American black walnut frame supports a plywood seat and backrest with veneers of oak, cherry and walnut. The design recalls traditional Japanese parquetry, interpreted in a modern form.

Tone Kids Table

Leif. designpark
L56cm x W49cm x H45cm

Tone 儿童桌由坚实的美国黑核桃木制成，以一种低调的现代形式展现出令人愉悦的拼花纹样，是 Tone 儿童椅的完美伴侣。

Tone Kids Table, made from solid planks of American black walnut, has an understated modern form that makes it a perfect accompaniment to the delectable parquetry of the Tone Kids Chair.

Toys & Crayons Shelf

ninetonine
Alberto Marcos for ninetonine
L150cm × W25cm × H27cm

这款书架包括一个线条优雅的木质抽屉，抽屉上带有两孔，便于暂时放置孩子们的玩具。旁边可以雕刻名字或词组，使整个书架变得更加独特。

Our shelf includes a practical drawer underneath its elegant line of wood, punctuated by pairs of different-sized holes, into which you can suspend baby's toys. Alongside, the option of a hand carved name or phrase can be carved into the wood to make it extra special.

Mouse

ninetonine
Alberto Marcos for ninetonine
L92.2cm x W50cm x H39.5cm

对于刚学步的孩子，这是一个有趣而伟大的设计，铅笔可以储存在 Mouse（老鼠）的木质耳朵里，以保持桌面整洁。三个螺丝钉固定桌腿，让拆卸变得非常简单。

A great, fun table for your toddler. Pencils can be conveniently stored inside the Mouse's wooden "ear", so things are kept neat and tidy, and made with three screw-on legs, assembly couldn't be easier.

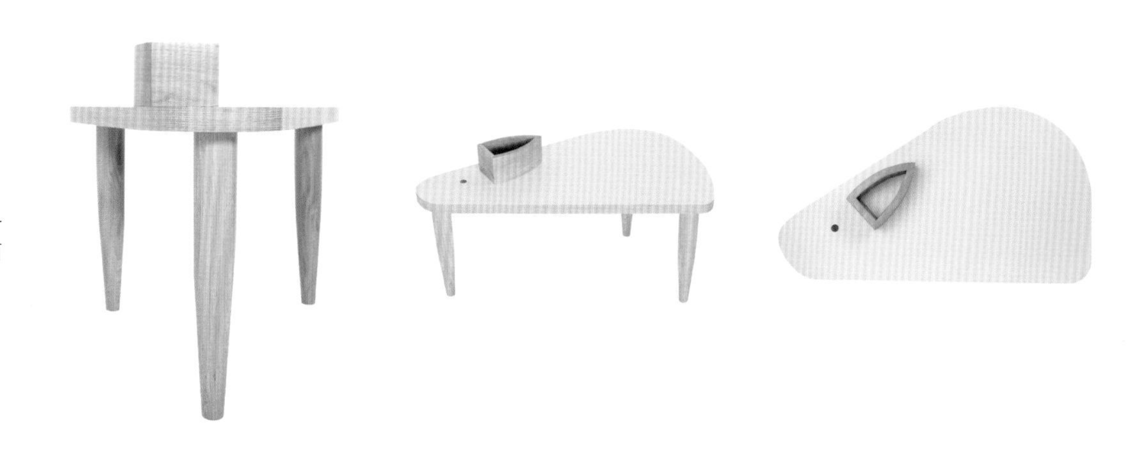

Shelf

ninetonine
Studio Merry for ninetonine
L50cm × W40cm × H115 cm

这款书架的高度方便孩子够取物品，且现代、实用，非常适合存放书和玩具，且书架会随着孩子的年龄增长始终发挥作用。

The ideal height for a child to reach all the shelves, this modern and practical shelf is a perfect solution for storage of your child's books and toys as he or she grows.

STOOL _01

ninetonine
Studio Merry for ninetonine
L25cm x W21cm x H21cm

STOOL 系列提供了两种尺寸的凳子，STOOL_01 是其中的微型版本。这个设计精悍实用，节约空间，可以尝试不同的色彩组合让整个空间变得更棒。

The collection offers two sizes of stool: this version is the mini Stool_01. Small but strong, you'll find you'll have space to fit a number of them in the nursery, perhaps trying out different colour combinations for a great result.

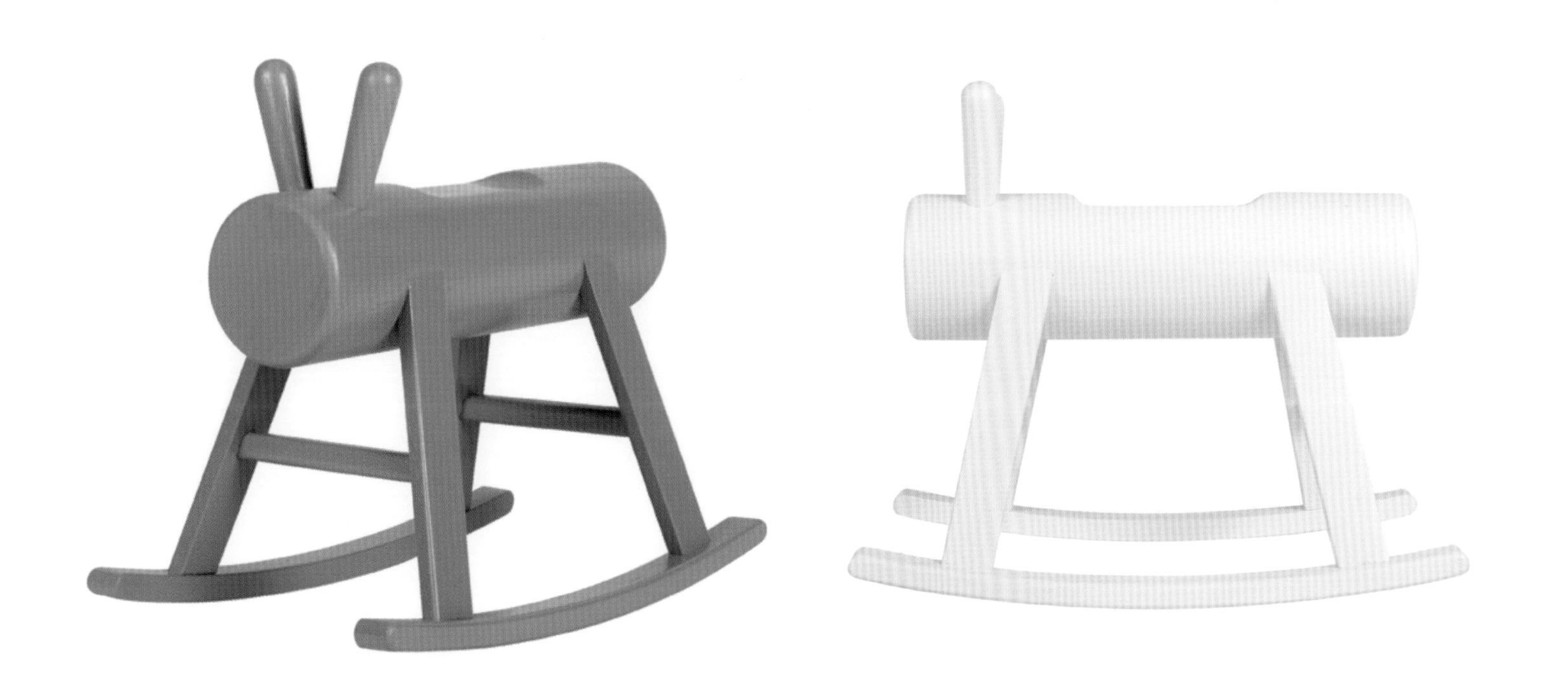

KITS Rocking Horse

ninetonine
Studio Merry for ninetonine
L49cm × W28cm × H41cm

一个精致的小摇马，两种颜色可供选择，轻但坚固，它会给您的孩子带来乐趣。

A design of wonderful rocking horse with two colors available, soft but strong, will bring your kids a lot of fun.

Bench

ninetonine
Studio Merry for ninetonine
L80cm x W26cm x H26cm

Bench 是 KITS 儿童桌的完美搭配，这个小板凳可以和该系列的许多产品任意组合，充分利用儿童房的每一寸空间。

The perfect pair to the KITS table, this little bench can also be combined with a few of the stools, to make the most of the space in your children's room.

Caterpillar

ninetonine
Alberto Marcos for ninetonine
L50cm × W40cm × H115cm

这个设计集凳子、游戏和玩具于一身。可一个一个单独使用，或拼成一列火车、一个长凳，以及任何你能想象到的东西。多种颜色可供选择。

A stool, game and toy all at the same time. You can use one individually, or join a number of them together to make a train, bench, or whatever your imagination desires! Available in different colours.

Robole

prodiz

一款外观造型完全取决于想象力的木质机器人。Robole 由一组简单的木质模块和磁铁组成。模块没有预设功能，可以任意组合创造出不同的机器人。关节的磁性设计使这款机器人不仅能组成稳定造型，而且能和用户进行良好的互动，可自由构建关节使其旋转。Robole 适合六岁以上孩子使用，也同时会给成年人带来无穷乐趣，专注培养想象力和逻辑力。波兰生产，涂无毒环保油漆。

Robole are wooden robots, whose appearance depends on the imagination of the holder. They come in a set of simple wooden blocks which are combined together by magnets. The parts have no function assigned, which makes it possible to interpret them freely, and create a different robot each time. Thanks to the magnetic joints, robole are not just stationary figures, but they may interact with the user. A user can construct their own joints that will enable changing robot's position and multiple-axis rotation of limbs. They are suitable for children from the age of 6, but they are a lot of fun for every adult as well. They develop the imagination and logical skills. They are produced locally in Poland and are painted with ecological non-toxic paints.

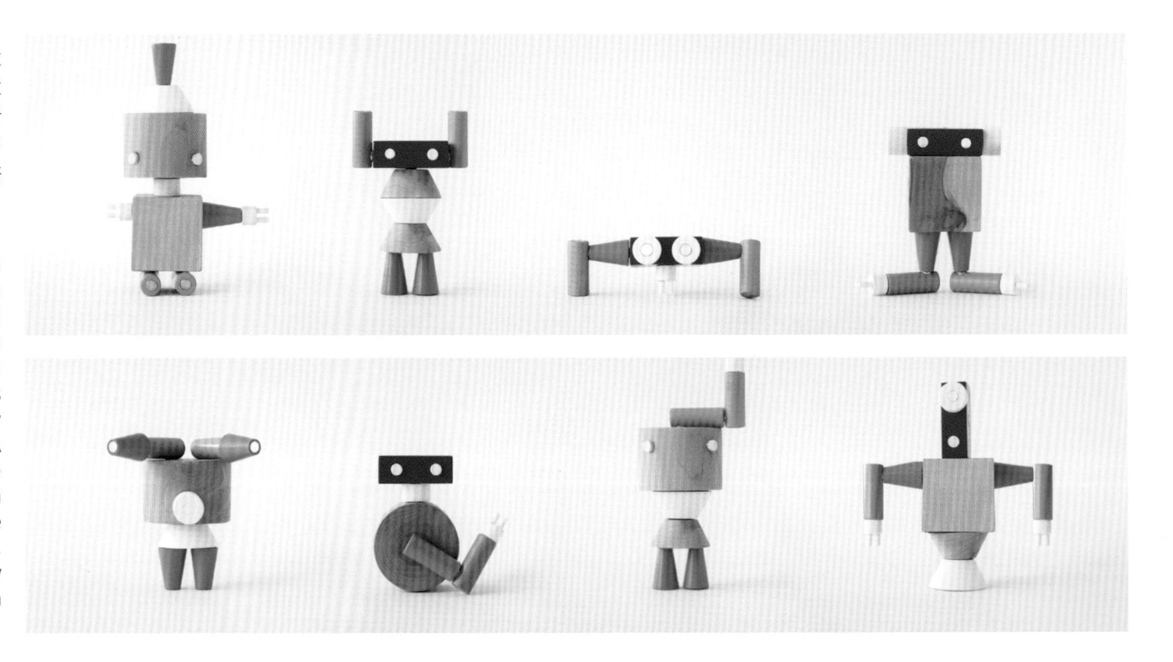

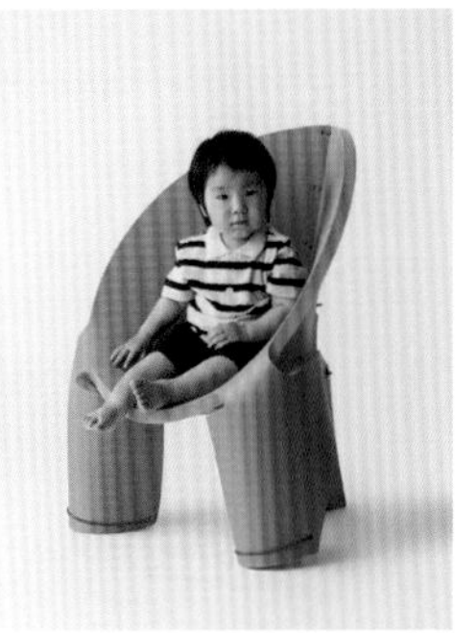

EVA Chair

h220430
L44cm × H80cm × D50cm

这款座椅通过将一块板子卷起来并用绳子固定来完成组装。

这款座椅能够在弯折后轻松恢复到平整状态，适合狭小空间储存，节约运输成本。

用于建造此款椅子的 EVA 材料重量轻，灵活度高，耐久性强，并且有多种颜色可供选择。若材料被意外吞食，也不会对人体造成伤害，非常适合儿童使用。

环保材料可高度回收利用，不含二恶英，非常适合我们未来的希望——孩子们使用。

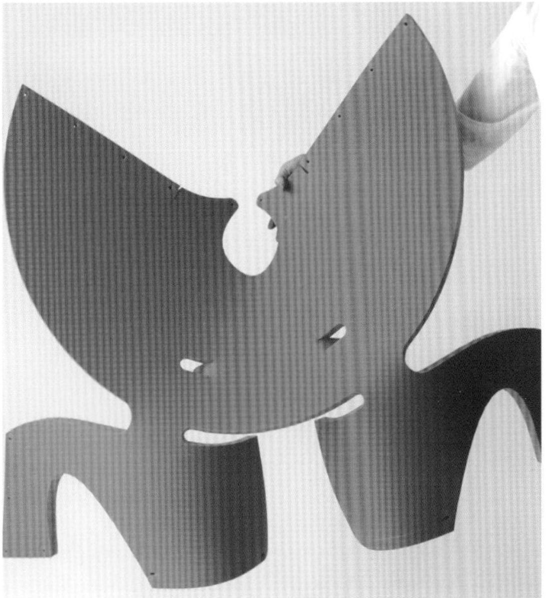

It can be completed just by rolling up a piece of board and fastening it with a string.

Because it can be returned to a planar shape easily, it can be stored even in small space. It also can save energy and cost at the time of shipment.

The material EVA that is used to build this chair is lightweight and holds a rich flexibility, superior durability, and various color variations. If the material accidentally enters into a mouth by any chance, it will be still safe so it is a good material for children.

In addition, with consideration of the material for high recyclability, no dioxin generation, and environmentally friendly, the material is suitable for children who play an important role in the future.

Plons

plons

Plons 专为孩子设计。不同于那种带彩色标签的小木质椅，Plons 设计独特，可以通过不同旋转随意改变造型。

旋转一次，Plons 就变成了一把平平的椅子，后背足以支撑个人重量，让你能舒舒服服地坐在桌边。再次转动，Plons 又会变成一个儿童躺卧椅。如果把后背延长，椅子向前倾斜，那么 Plons 又成为一个舒服的休闲椅，供玩游戏或看电视时使用。

再次翻转，Plons 就会变成一个能够建造高塔的大玩具。有一种旋转甚至能让它变身为一个高大的机器人。

Plons 多种多样的玩法给孩子提供了无限挑战，同时发展他们强烈的空间认知能力。

Plons 有浅绿色、蓝色、粉色和橘黄色四种颜色可供选择，室内室外均可以使用。宽大有力的支撑结构使 Plons 同样适用于海滩沙地。

Plons 重四公斤，由一块坚固、可循环利用的无公害的聚乙烯制成，可以轻松抵御最强烈的的阳光照射。

Plons 性能安全，已被授予 CE 奖章，通过了欧洲防火安全水平测试，不含任何对人体有害物质。

Plons is a children's chair. Not one of those small, wooden chairs with a colorful sticker on it. Plons is unique. With every twist you change what Plons is.

With one twist, Plons gets a flat seat and a supportive back for comfort seating at a table. Twist again to turn Plons into the very first children's lounge chair. The back is longer and the seat leans backwards to make Plons the ultimate relax chair for gaming or watching TV.

Turned over once more, Plons becomes a big toy fit to build tall towers with. There's even a turn which transforms Plons into a bright fellow with strechted arms.

The multiple uses of Plons challenge a child to develop a strong spacial awareness.

Plons is available in bright green, blue, pink and orange. Plons can be used inside as well as out of the house. Thanks to having wide supportive legs, Plons won't even sink in the fine sand at the beach.

Plons weighs four kilograms and is made out of one solid piece of recyclable, harmless polythyleen which can easily withstand the brightest sunlight.

Plons is safe. It has been awarded the CE seal and complies with European standards for fire safety and does not contain any hazardous substances.

Molecola 3

PLAY+
Design M+A+P Designstudio

Caleidoscopio / MirrorTriangle

PLAY+

Design Tullio Zini Architetto

Nido / Cocoon

PLAY+
Design Terri Pecora

Atollo / Atoll

PLAY+
Design ZPZ Partners

Cactus

PLAY+
Design Donegani & Lauda

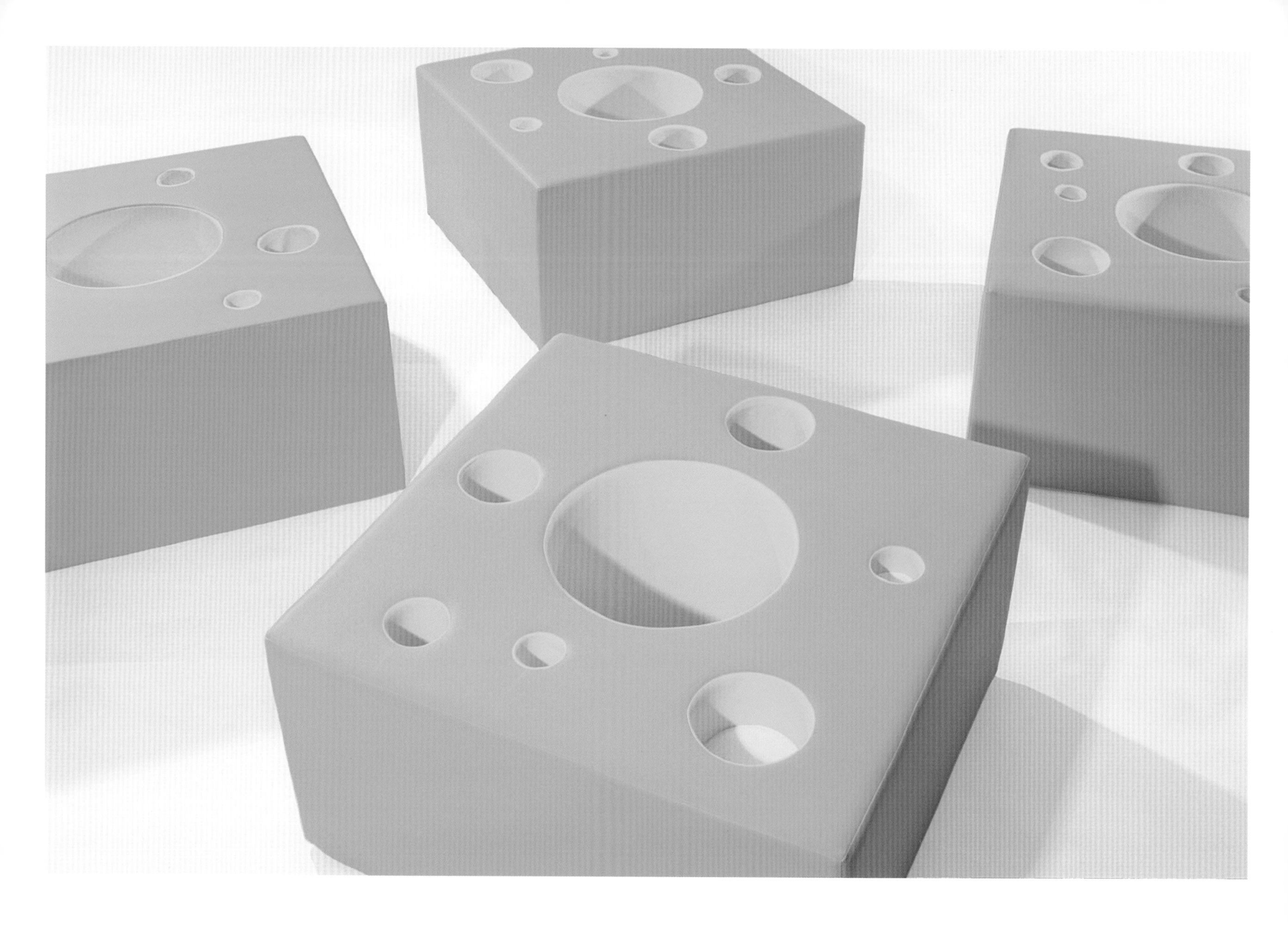

Buchi / Holes

PLAY+
Design Andrea Branzi

Intrecci / Weaves

PLAY+
Design Total Tool

Navicella / Lem

PLAY+
Design ZPZ Partners

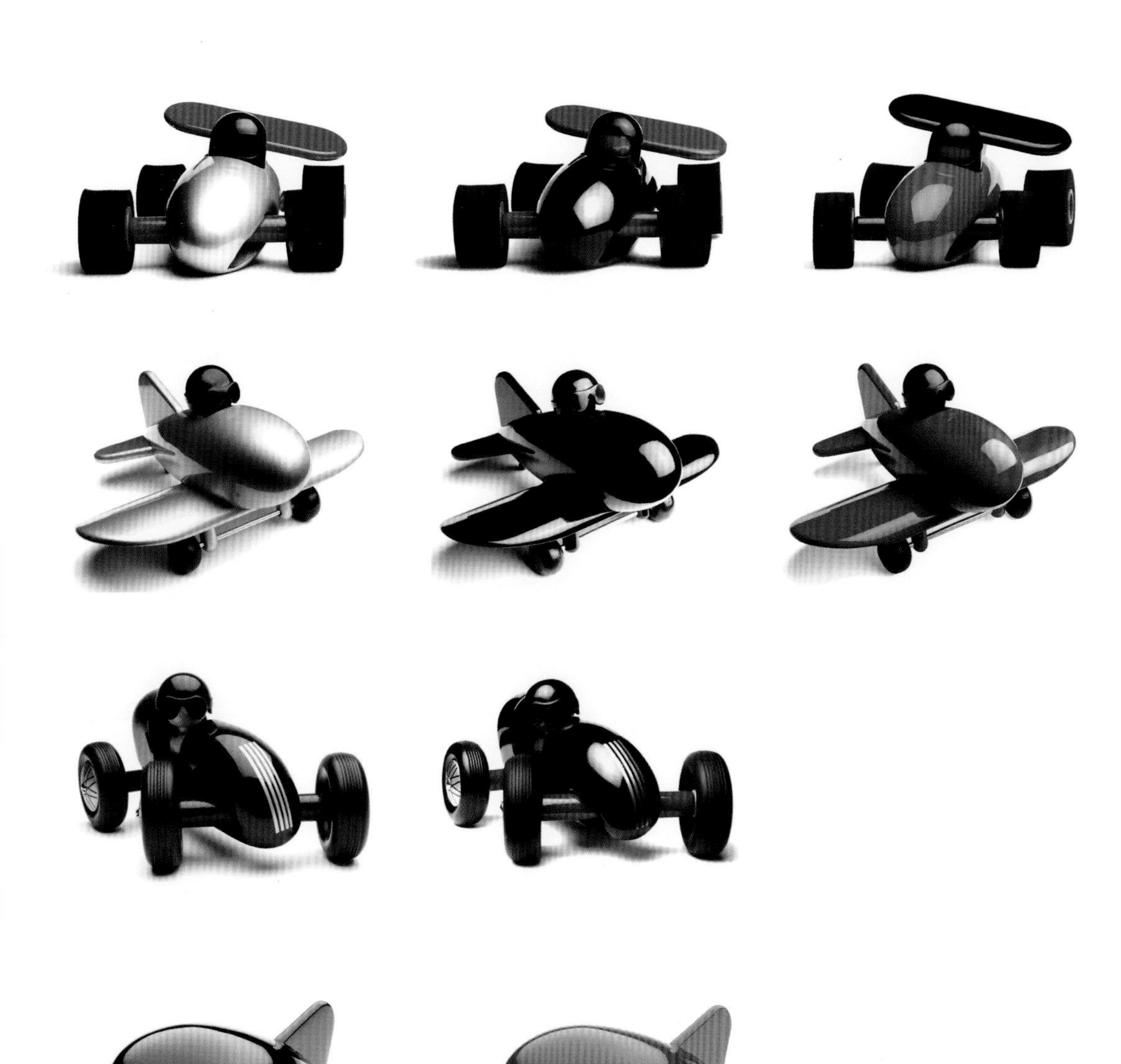

Racer F1

Playsam

瑞典优秀设计奖获奖作品 Racer F1 外表色泽光彩夺目，形体线条令人愉悦，是一件为全世界赛车爱好者准备的特别礼物，这款坚实的木质赛车给人一种全新的感觉，仿佛在你身边“呜！呜！”地呼啸而过。

Named an Excellent Swedish Design recipient, the Racer F1 flaunts a high-luster exterior and an exhilerating body design. An exclusive gift for racing fans around the world, this solid wooden race car gives a whole new feeling to the expression "vroom, vroom."

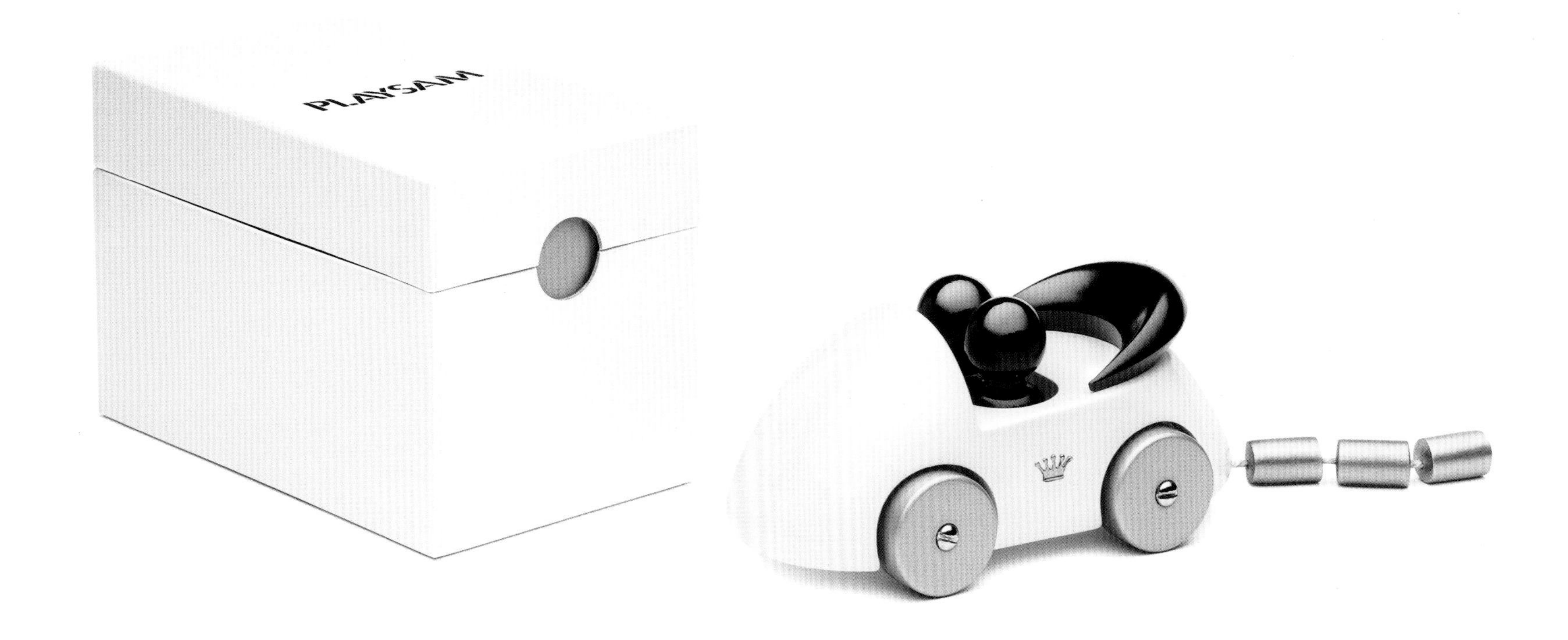

Royal Wedding Car

Playsam

Playsam 皇家婚礼汽车是一款为婚礼准备的时尚礼品，一个向王储维多利亚公主和丹尼尔的婚姻致敬的产品。

Playsam Royal Wedding Car is a fashion gift for wedding. A homage for Crown Princess Victoria & Daniel marriage.

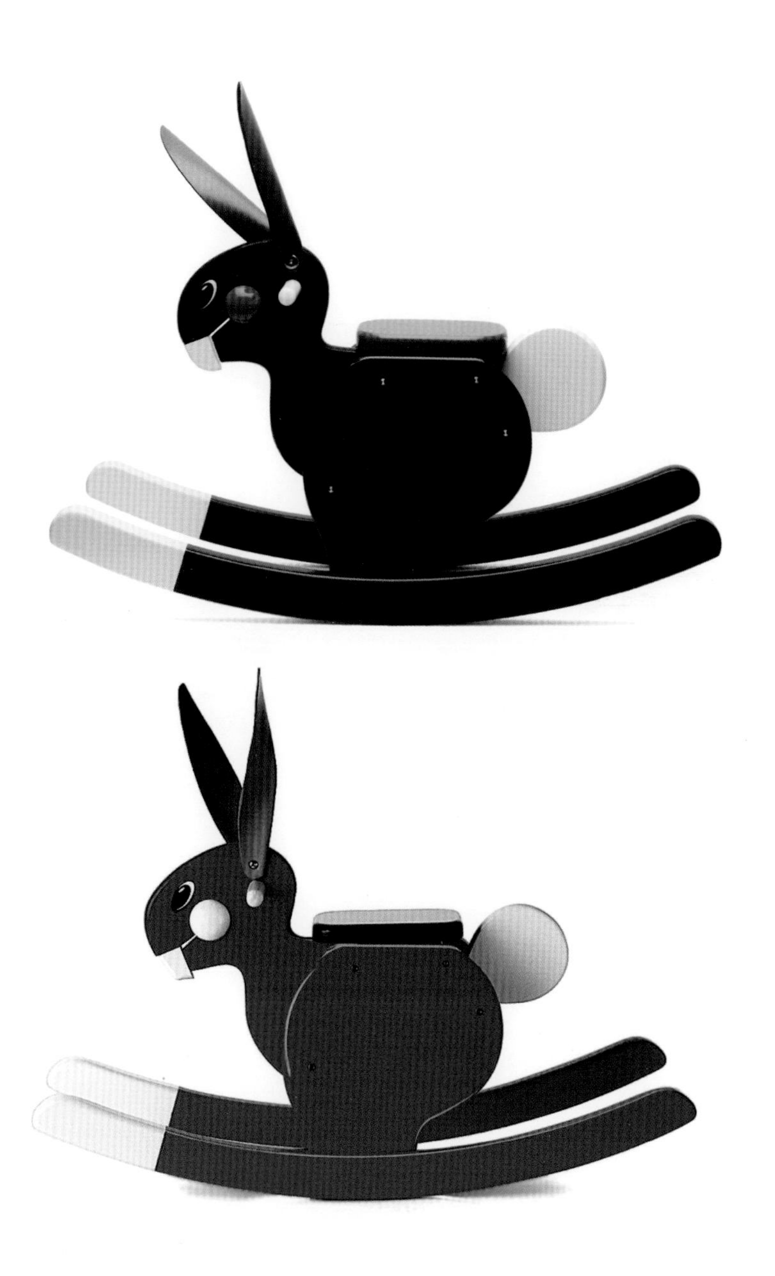

Rocking Rabbit

Playsam

如果要参加传统摇摆木马比赛，这只摇摆兔肯定能拔得头筹！创造性的设计和夺目光彩让这只兔子成为瑞典优秀设计奖的宠儿！

A playful rendition of the traditional rocking horse, this rocking rabbit is sure to win the race! Its creative design and glossy shine have earned this long-eared critter the Excellent Swedish Design seal of approval!

Streamliner Classic Organic

Playsam

这是一款经典的有机木质玩具，以其独特风格和时髦外观被瑞典国立博物馆收藏，一直是孩子和大人的至爱。

A classic organic Playsam design, the Streamliner Classic Car has been selected as Swedish Design Classic by the Swedish National Museum for its inventive style and sleek surface. It's no wonder that this organic wooden car has also been a long-standing favorite for children both big and small.

Nautilus

Alicia Bastian
121cm×170cm

Nautilus 的设计灵感源于“活到老，学到老”的生活理念。这件传家宝中可以放置家中最有价值的书籍。上升的形态代表读者年龄的增长，也象征着在以后的学习生涯中将要获得的无尽知识。当其成线形形状时，可以用来分隔空间。闭合时，Nautilus 的环形结构营造出一个舒适的阅读空间。

Nautilus is a bookshelf inspired by the lifelong need for growth through learning. As an heirloom object, it contains only the family's most valued books. The ascending pattern represents the growth of the reader's age as well as the knowledge the reader will obtain over time in their pursuit of lifelong learning. When in linear form, Nautilus can be utilized as a room divider. When closed, Nautilus's circular form becomes an adapted reading space.

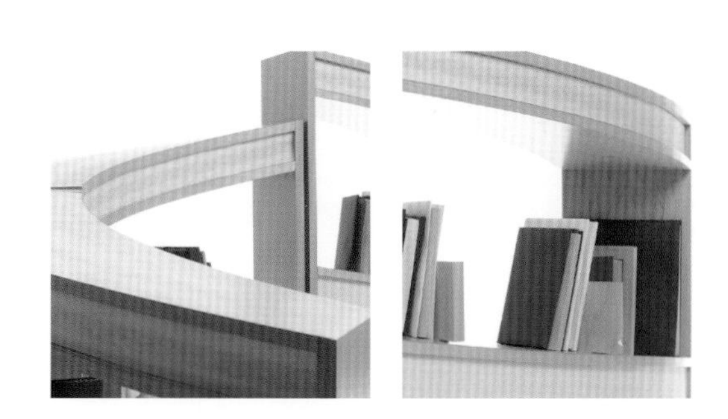

Boattail Racer

Auditorium Toy Company

这是一款将形式美学和空气动力学完美结合的佳作，灵感源自 20 世纪早期的流线型和低阻力汽车设计。Boattail 赛车由耐腐蚀材料手工制作而成。

它的木质车身以有着 13 层崎岖花纹的海桦木雕刻而成，尊贵和强健的造型来自于 77.5 毫米玛瑙核心工业级轮毂，可以在精密轴承的作用下轻松悬浮。

保护 Boattail 赛车的是以四层抗酸博物馆板材手工打造而成的定制储存箱。

A masterpiece of aesthetic form and function inspired by the slippery lines of early 20th century low-drag coachwork — designs formulated by the science of aerodynamics, yet strikingly gorgeous. The Boattail Racer is handmade from materials that time cannot easily erode.

The wooden bodywork is sculpted from rugged 13 ply Baltic birch. The muscular stance comes courtesy of the 77.5 millimeter carnelian-core industrial grade wheels, which float effortlessly on precision bearings. Solid stainless steel axles and fasteners ensure faithful service for generations to come.

The Boattail Racer is protected by a custom archive box handcrafted from black-core 4 ply acid-free museum board.

The Committee

Auditorium Toy Company

六个商人模样的人物在同一辆车中进行轻松的旅行，其中一名乘客似乎受到限制。车厢中有许多设备和工具（包括长绳、录音机、棒球棍、铲子等等）。这个场景中的每个元素单独看起来并无任何凶兆，但当他们组合在一起却危险重重。

The Committee 通过集体智慧，解释了这一驾驶过程中的潜在危险，在这个集体中成员们心血来潮的决定已经超越了最初的目的和动机。

在这无尽的驾驶过程中你才是最重要的驾驶者。

由工业级材料手工制造的 25 个产品，有各自编号和认证文件。

Six business-attired figures, viewed from above in diagrammatic form, are passengers in a vehicle on a relaxing outing. However, one of the passengers is restrained. The trunk reveals various instruments and tools (length of rope, tape recorder, baseball bat, shovel, etc.). While each element in this scenario is individually innocuous, together they make up an undeniably sinister collection.

"The Committee" exposes the risk to the potential of an idea when driven by a collective will — when the original motivation and criteria take a back seat to the impulsive decisions of individual committee members.

You are the ultimate driver in this unfolding narrative of control.

Meticulously handcrafted from industrial grade materials in a limited production of 25 handmade, numbered and certified pieces.

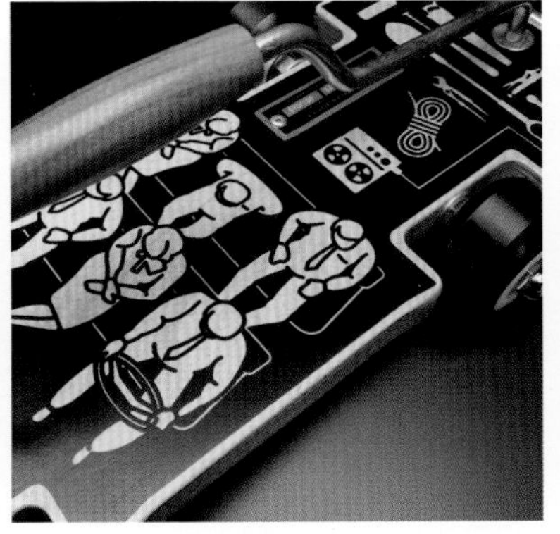

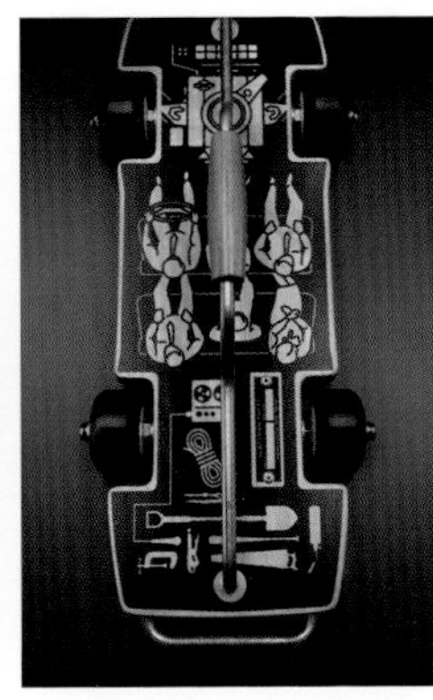

Wishbone Bike

Wishbone Design Studio

Wishbone 自行车是一个三合一的玩具。根据可持续性设计原则，它既可以是婴儿学步车，也可以是室内用三轮车和平衡车，三位一体。

孩子一岁时，可当做三轮车使用。孩子长到三岁时，可以把它转换成两轮的平衡车。再后来这辆车可以发挥更大作用，一直伴随孩子到五岁。

Wishbone 是婴儿学步、幼童骑行、学前儿童轻松掌握平衡的理想选择。

由新种植的桦木和桉木制成，可循环使用的有机帆布袋子包装。Wishbone 设计工作室是一个零碳认证机构。

Wishbone Bike is a 3in1 convertible ride-on toy. Based on a sustainable design philosophy, Wishbone Bike is a baby walker, indoor ride-on toy, tricycle and balance bike all-in-one.

It starts at age one with three wheels, converts at age 3 to a two wheel balance bike, then later the "wishbone" frame inverts, so the bike grows with a child from 12 months up to 5 years of age.

Babies learn to walk. Toddlers learn to steer, coast and control their speed instinctively. Pre-schoolers learn to balance with ease.

Made from reforested birch and eucalyptus, recycled packaging and with an organic canvas axle bag. Wishbone Design Studio is a certified CarboNZero organisation.

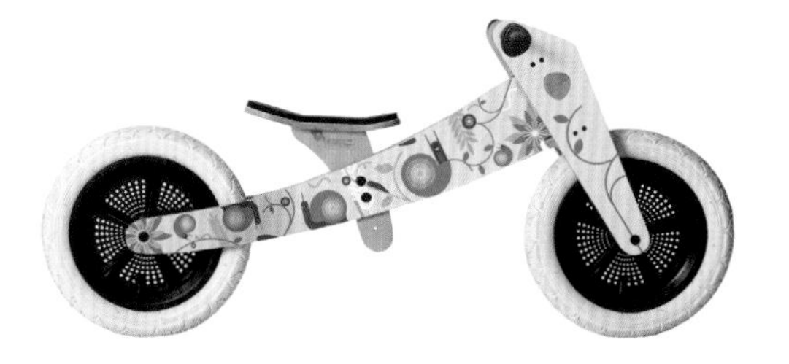

Compas Desk

Atelier Charivari

来自 20 世纪 50 年代的 Compas 书桌表面涂苔藓绿色。

Compas Desk from the 1950's painted in a lichen green.

Violet Compas Desk

Atelier Charivari

这是一款罕见的紫罗兰色和木炭色的 Compas 书桌。

Rare compas desk in Violet and Charcoal.

School Desk

Atelier Charivari

20 世纪 60 年代的工业椅，曾用作教师书桌。

Once was the Teacher's desk, from the 1960's, industrial chair.

M-shaped Chair

ASAI

M 形的椅子，适用于 3 岁儿童，具有防水、耐擦洗、防静电的优点。

A chair for child of 3 years old which is waterproof, scour bearing and antistatic.

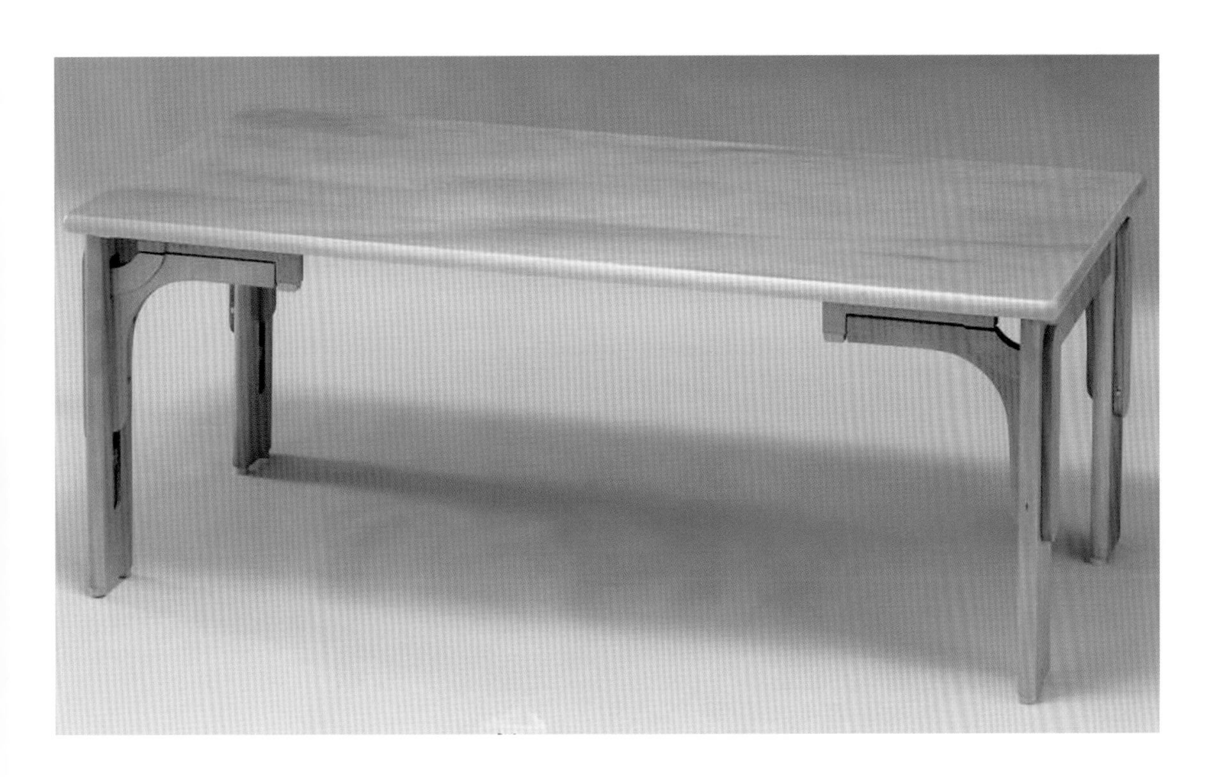

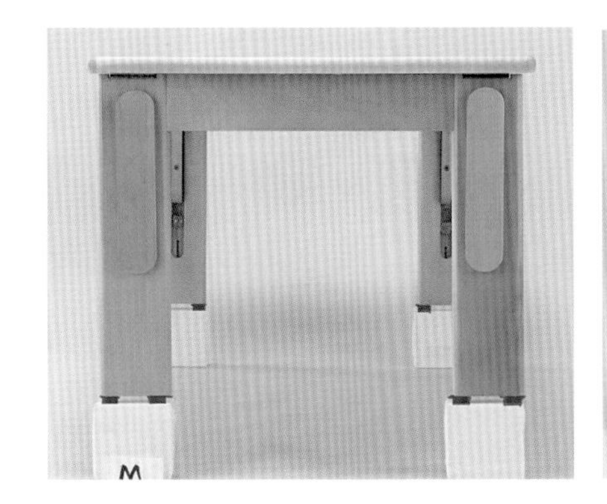

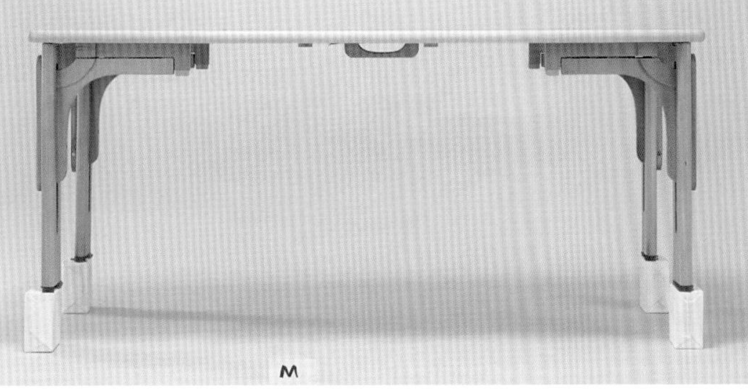

M-shaped Table

ASAI

M 形的桌椅适用于 3 岁儿童，具有防水、耐擦洗、防静电的优点。

An M-shaped desk made for children of three years old, which is waterproof, scour bearing and antistatic.

L-shaped Desk for Two

ASAI

L 形的桌子适用于 5 岁儿童，具有防水、耐擦洗、防静电的优点。

An L-shaped desk made for children of five years old. Be waterproof, scour bearing and antistatic.

L-shaped Desk for One

ASAI

L 形的桌椅适用于 3 岁儿童，具有防水、耐擦洗、防静电的优点。

An L-shaped desk made for children of five years old. Be waterproof, scour bearing and antistatic.

Stepladder Chair

if–j
W29cm × D37.75cm × H37cm

对曾以 CodomonoCoto 品牌发表的脚凳椅进行了量产化重新设计的椅子产品。

既作为儿童用的椅子，也可以作为儿童在洗脸刷牙和在厨房帮忙时用的脚凳。在孩子长大不用后还可以作为家庭的脚凳使用。

A stepladder chair of new design based on CodomonoCoto with mass production.

A chair for child to sit and a stool for them to stand when washing in bathroom and helping in the mass kitchen.

Baby in Table

if–j
W110cm × D40cm × H37cm

Nesting Combination

if-j
W37cm × D28cm × H27cm
W54cm × D40cm × H42cm
W70cm × D48.5cm × H71cm

这个 3 件套桌可以组合使用，嵌套放置时占用空间小。小桌可供 4 岁以下的儿童使用。中桌加小桌时，小桌可当做凳子使用，供小学低年级的儿童使用。大桌加中桌，中桌可当做凳子使用，供小学以上的孩子使用。

在孩子不再使用后，小桌可作为玄关处的穿鞋凳，中桌可作为沙发的侧桌，大桌可作为起居室的工作桌等继续使用。

由于桌子下方设有搁脚的踏板，所以脚够不到地面也没有问题。

中桌可当做凳子使用，较宽的凳面可与妈妈一起坐下学习。

The nesting combination consists of three tables that save a lot of space. The small table is made for children of 4 years old or younger. The medium size can be used in combination with the small one for lower grade students or younger, with the small table as a stool. When the big one is combined with the medium size, the latter becomes a stool, a combination for children beyond elementary school.

When the child grows up, the combination also has many applications, with the small table as a bench in the vestibule, the medium one besides the sofa and the big one a workbench in living room.

A comfortable foot step for leg supporting under the table is specially designed for those whose feet cannot reach the ground.

Table of medium size with a wider surface allows mother to sit with the child.

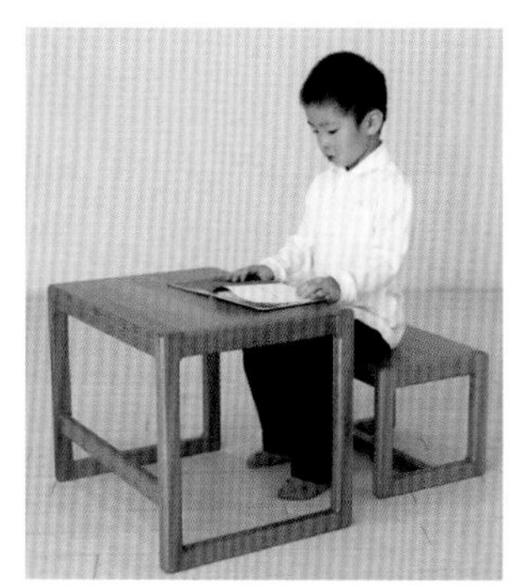

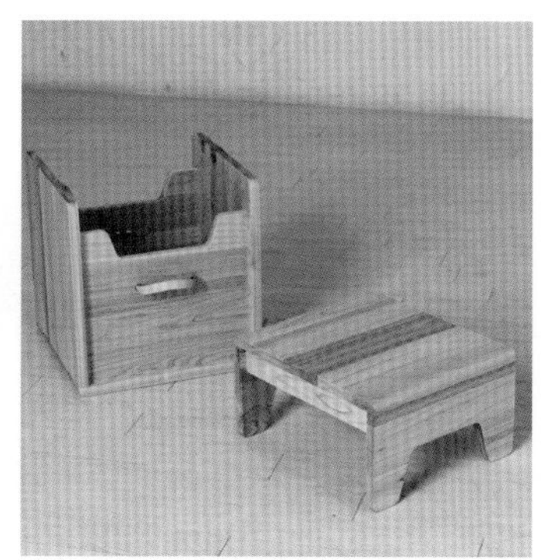

Box Desk

if–j
Body dimension: L38cm × W38cm × H39cm
Lid dimension: L38cm × W34cm × H20cm

箱盖装上时可作为玩具箱使用。盖子也可以当做小孩子们玩摆家家时用的小桌子。盖子装上时也可以作为凳子使用。带有真皮把手。

Covered with the lid, it becomes a toy box .The lid turns into a short desk for playing children, and a stool when covering the case again. With leather handle.

Birds Tree

Kidsonroof
H85cm × L70cm × W70cm

和树上的15只小鸟一起玩耍吧！这棵树不仅仅是儿童房的完美装饰品，也是孩子们有趣的玩具！

小鸟们有的绕树而飞，有的停在树上或者地上。在树枝上的鸟巢里，有一只嗷嗷待哺的小雏鸟，雏鸟周围有家燕、翠鸟、喜鹊、猫头鹰、麻雀、鹦鹉、灰色的鹡鸰、蓝色的山雀、吃腐肉的乌鸦、欧洲知更鸟，还有一只身着点点斑纹的啄木鸟。这款玩具适合三岁以上儿童，有七十个部件，由可回收纸板制成。

Play with this tree and its 15 birds ! This tree is not only a wonderful piece of decoration for the children's room...it's a wonderful play object for the kids!

Have the birds fly around, land again in the tree or hop on the ground. In the branches is a nest with a hungry baby bird and surrounding him you'll find a barn swallow, a kingfisher, a magpie, a nuthatch, an owl, a sparrow, a grey wagtail, a blue tit, a carrion crow, an European robin and a lesser spotted woodpecker. Age 3+. 70 pieces. Made of recycled cardboard.

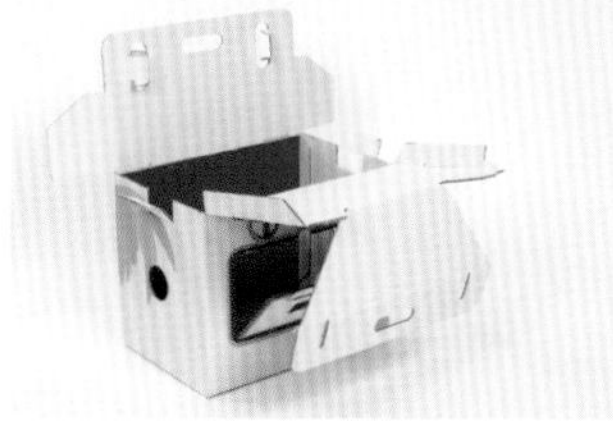

Cococrico Cooker

Kidsonroof
D22cm × L34cm × H45cm

一个专为初级厨师准备的纸板厨具。轻质而小巧，但足够让孩子与父母在厨房中体验亲自下厨的乐趣。可以放在桌上、地板上，以及任何能够放得下罐子或者深平底锅的地方。总之，尽情发挥吧！这款产品由可回收纸板制成。

A cardboard cooker for the beginning cook. Lightweight and small... though big enough to start cooking with papa or mama in the kitchen. To put on the table, on the ground or anywhere where the little cook installs his cooking pots or saucepans. And ...cook a doodle do said the rooster at the back of the cooker... (cocorico). Made of recycled cardboard.

Birds Tree

Kidsonroof

Volunme Zero是Kidsonroof系列家具组合中的一款，由实心榉木制成，可以变为平板形状，无须任何螺丝或钉子即可组装。

该系列包括一把椅子、一张长凳和一张桌子。三个部件都有红、绿两种颜色可供选择。

Volume Zero is the Kidsonroof furniture collection. Made of solid beech wood this collection comes flatpacked and is very easy to set up without any screws or nails.

This collection consists of a chair, a bench and a table. All three are available in green or in red.

Casa Cabana

Kidsonroof
85cm × 70cm × 70cm

Casa Cabana 小屋有两种款式，一款为通体白色，另一款上面画有一只小猫蹲在树上、一只小蚂蚁、一只小兔子和一只小松鼠。

两种款式都提供了充足的空间用来上色、涂抹或者装饰，从而创造属于每个孩子自己的 Casa Cabana 小屋，让它成为孩子们的树屋、豪宅或者藏身屋吧。你想给它取几个名字就取几个名字。如果想单独待一会儿，就在门口摆上“请勿入内”的标识。Casa Cabana 小屋有一扇门、七个窗户和多个观察孔。由可回收纸板和生物可降解材料制成。

Casa Cabana comes either completely white or with a cat on a tree, a small ant, a rabbit and a squirrel.

In any case more than enough room to color, paint and decorate your very own Casa Cabana. Let it be your tree house, mansion or hiding shed. Give it as many names as you like... and do you want to be left alone a bit? ... put a sign "no trespassing". Casa Cabana has a door, seven windows and many spy holes. Made of recycled paper and biodegradable.

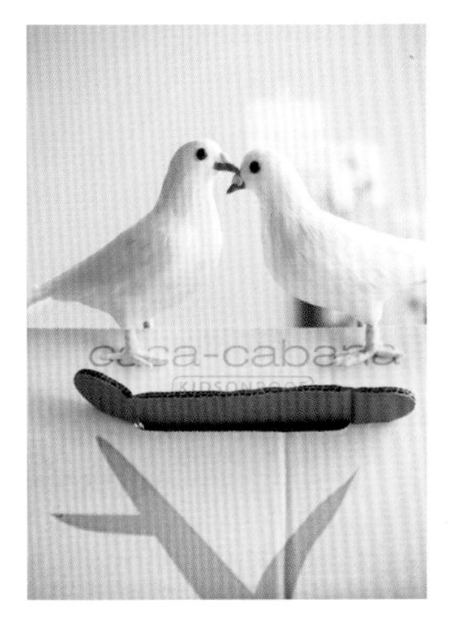

Rocket

Kidsonroof
140cm×100cm×100cm

首个为未来的飞行员们准备的模拟训练场。

The First trainings module for all future astronauts.

Totem Deer

Kidsonroof
22cm×35cm

快来建造这个美丽的森林小鹿，进入它那个大树参天直通天际的奇妙世界吧！

由可回收纸板制成，适合六岁以上儿童，共42个部件。

Build this beautiful forest deer! Roam into their world where trees touch the sky.

Made of recycled cardboard. Age 6+, 42 building pieces.

Bergdorf

Kiri Martin

Bergdorf 是一个小猪造型的储蓄罐，创意新颖，原料可循环使用。这个纸浆模塑小猪不仅外观可爱，且能够让孩子们理解金钱和储蓄的意义。尽管使用材料不多，储蓄罐依然可以装满硬币，平稳站立。如果孩子们要取出硬币，可撕开储蓄罐，并循环再利用。所选材料提供了多种客户定制选项。孩子可以在上面画画，让自己的储蓄罐拥有非凡个性。这款储蓄罐也提供了不同颜色和纸质选择。孩子们需要硬币时，可以体验到破开储蓄罐而拿到硬币的传统习惯，整个过程完全安全、清洁、可持续。孩子们定会乐在其中。

Bergdorf is an innovative and sustainable rendition of the iconic piggy bank. This molded paper pulp piggy bank is not only adorable, it also teaches kids the value of money and saving. Using minimal materials, the bank is stable enough to fill with coins but can still be ripped apart and then recycled when its time for kids to retrieve their savings. The chosen material provides endless options for customization. Kids can draw on the bank to give it a unique character and personality. The banks can also be made in any color as well as varying paper textures. This toy brings back the ritual of breaking your bank when its time to retrieve your coins, but offers a completely safe, clean and sustainable option which any kid could appreciate and enjoy.

Blue

Kidsonroof

用超级加固纸板制成的三维建筑，儿童可进行再次创作，由可回收纸板制成。

A three dimensional building with super sturdy cardboard, allowing children a chance of re-creation. Made of recycled cardboard.

STOOLESK

COLLECT Furniture

STOOLESK 是一款有两种大小的双功能凳子。年龄在一至五岁的小孩子可以坐在较小凳子上把较大的凳子当做桌子。五岁以上的孩子和成年人则可以将它用作板凳、床头桌或者咖啡桌。

STOOLESK 和 STOOL 因其完全相同的角度和良好的稳定性而能够完美地组合在一起。STOOLESK 的一面有存放纸和铅笔的空间，当你的创造灵感突发之时，总能方便地拿到纸笔来记录灵感的火花。

由弗雷德里克・柯莱特（Frederic Collette）设计，采用厚重的橡木板制成，遍体覆有漂亮的木纹。

产品已达到北欧天鹅生态标志系统（Nordic ECO label SWAN）要求，证明其具有良好的可持续性、耐用性及安全性。

STOOLESK is basically STOOL in double size with double functions: Small children aged 1-5 may use it as a desk, when sitting on STOOL. Children older than 5 years and adults may use it as a bench, bed side table or coffee table.

STOOLESK and STOOL fit perfectly together having the same angles and good stability. STOOLESK has room for paper and pencils on one side – always at hand, when creativity calls.

STOOLESK is made of wide massive oak planks, to fully enjoy the wood grain. Designed by Frederic Collette.

STOOLESK has achieved the Nordic ECOlabel SWAN, documenting its durability, health safety and environmentally sound production.

A CHAIR

COLLECT Furniture

这款形象家具是弗雷德里克·柯莱特专为三至十岁儿童设计的 A 系列中的一员。该系列家具的设计灵感来源于字母 A 的斜边。

A 字造型的特殊角度令椅身稳定，不易被儿童翘起，并能被优雅地折叠起来。不仅如此，这个角度还使得座位下可安放抽屉，供孩子们放置玩具和其他物品。更重要的是，挪动椅子时，抽屉的磁性设计保证它不会被轻易拉开。

This very graphic furniture is part of our A-series designed by Frederic Collette for children 3-10 years old. The A-series is built up around the inclined sides of the letter A.

These angles give stability, so that children cannot easily tilt the furniture. But these angles also enable A CHAIR to be stacked in an elegant way. At the same time, this gives room for a drawer below the seat, where children can put toys and things. The stacked chairs form a practical chest of drawers. Furthermore, the drawers are equipped with magnets, so that they do not open while moving A CHAIR.

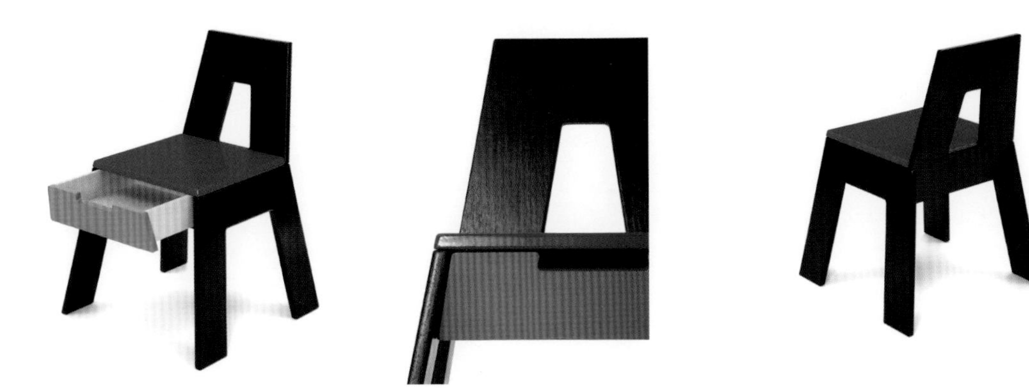

HANG A TABLE

COLLECT Furniture

这款可以挂在墙上的桌子适合较小卧室中的一个孩子使用。桌子设有三个抽屉，可放置纸笔。随着孩子渐渐长大，还可以将桌子挂得更高，也可以用作梳妆台，或放在玄关用来保存钥匙、信件。这款桌子可以与该系列的椅子完美搭配。

由弗雷德里克·柯莱特为三至十岁儿童设计。年龄稍大的儿童或成人也可以使用。

HANG A TABLE is wall suspended and designed for one child in a small bedroom. HANG A TABLE has 3 drawers for paper and pencils. Hang it higher as children grow or use it as beauty desk or hallway table with storage for keys and mail. It fits perfectly with A CHAIR.

It is designed by Frederic Collette for children 3-10 years old with A CHAIR and can be used also by bigger children or adults.

A TABLE

COLLECT Furniture

这款形象家具是弗雷德里克·柯莱特专为三至十岁儿童设计的 A 系列中的一员。

桌子可供两个孩子相对而坐。桌子设有六个抽屉（每侧有三个），可放置纸张、铅笔、蜡笔等。它也是 A 系列椅子的完美搭档，这组家具曾在 2006 年赢得 Formland 设计大奖。

This very graphic piece of furniture is part of our A-series designed by Frederic Collette for children 3-10 years old.

A TABLE is designed for 2 children sitting opposite each other and has 6 drawers (3 on each side) for paper, pencils, crayons and stuff. A TABLE fits perfectly together with A CHAIR and these were nominated as a set for the Formland design price in 2006.

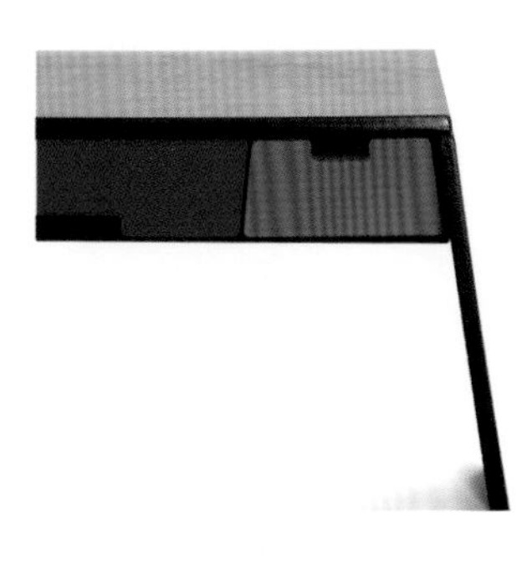

Children´s Book Chair

Charlotte Friis

这款椅子适用于一至六岁的孩子，通过书本翻页可以调整座椅高度。孩子们可以在书页后侧插入自己的图画或者照片,同时也可以把这些当做椅子的靠背。当“我的故事”，的所有书页都插满自己的图片时，这本书就会变得个性十足。图片松散地插在椅子上，可以随时更新。座位下面是一个私密空间，孩子们可以把泰迪熊或者其他喜欢的玩具藏在那里。书本合上时就变成了一个小桌子。

The chair suits children from 1 to 6 years, and the seating height can be adjusted by turning pages in the book. The Children can insert their own drawings and/or photographs on the back side of the pages, which also works as the back rest of the chair. The book holds the title "my Story", because the book becomes very personal, when all the pages are filled with images. The images are loosely attached to the chair and can easily be updated. Underneath the seats is a secret space, where the child can hide a teddy or another favorite toy. When the book is closed it becomes a small table.

AKIYUKI SASAKI

AKIYUKI SASAKI DESIGN

如同打开一大片纸，将其折成桌子和椅子的形状一样，这款产品的每一个部件都十分简易。设计师想让它们尽可能地抽象、简洁，与日本手工艺传统紧密结合起来。为了达到理想的形状，设计师采用了模制胶合板技术。想象一下，一群孩子围在我们放置大块板材或轻质木材的桌子旁边，这些可以使孩子们真正体会到这种构思——他们会开始进行概念化的思考，思考出一些全新的东西，一些设计师自己从未想象到的东西。它可以叠为三层，虽然乍看上去可能没有实际用处，但它为孩子们创造了一种环境，可供他们随意摆弄，使孩子们的创造力最大限度地发挥出来。设计师选择了山毛榉木作为主要材料，因为我希望孩子们能够感受并且爱上自然材料本身。

Each item is as if you simply folded a massive piece of paper to give it a shape of a table and a chair. I thought of keeping them as abstract as possible, simplicity so well renowned and attached to the artisan traditions of Japan. In order to achieve his idea and shape the designer adopted the technology of molding plywood. Imagine a gather of a few kids around the tables where we place big flat sheets or balsa for kids to actually learn how this was conceived — they might end up conceptualizing something totally new, something the designer would never have imagined himself. It can be stacked up to 3 layers — though at first sight of the product may, appear functionally void — it creates an environment where kids can maximize their own creativity and mess around as much as they please. The designer selected beech wood as main material because he hopes kids feel and love the natural material itself.

FRAME TABLE

COLLECT Furniture

框架系列的最新产品——框架型桌子面世了。

整个框架系列产品均由弗雷德里克·柯莱特设计——厚重的橡胶木材料结合经典连接设计而成的矩形框架型家具。

框架型桌子为一至五岁孩子设计，让他们能够面对面坐着。桌子表面层压塑料装饰板，可用作黑板，供孩子在上面绘画书写。下方为一个搁架，可从桌子两边放置粉笔、纸张及其他物品。

孩子长大后，框架型桌子可作咖啡桌用，下方的搁架用来放置杂志及电视遥控器。

The new member in the new FRAME-series, FRAME TABLE is now available.

The FRAME-series is designed by Frederic Collette and consists of furniture built up of rectangular frames in massive oak-wood assembled with classic joints.

FRAME TABLE is designed for 1-5 year-old children sitting opposite each other. The table top is laminated with strong FORMICA, which can be used as a chalkboard plate children can draw upon directly. Below is a shelf on either side of the table with room for chalk, paper and things.

When children grow up, FRAME TABLE can be used as a coffee table for magazines and remote on the shelf below.

FRAME CHAIR

COLLECT Furniture

框架型椅子是框架系列的第一款家具产品。本系列产品都是由厚重的橡胶木材料结合经典的连接设计组合而成的矩形家具。

这款产品由四个相互铆合的框架组成，这个结构支撑起了帆布的座位和靠背。帆布采用百分之百经认证的有机棉花制成，可拆洗。

坐在框架型椅子上看电视非常舒服。它也很适合孩子攀爬。

由弗雷德里克·柯莱特为一至五岁的孩子专门设计。

框架型椅子达到了北欧天鹅生态标志系统的要求，具有可持续性、耐用性及安全性的优点。

First member in the FRAME-series consisting of furniture built up of rectangular frames in massive oak-wood assembled with classic joints.

FRAME CHAIR is composed of 4 frames fixed onto each other to form the supporting structure for the canvas seat and back. This canvas is 100% certified organic cotton, detachable and washable.

FRAME CHAIR is a favorite place to watch TV, but is also good for climbing!

Designed by Frederic Collette for children 1-5 years old.

FRAME CHAIR has achieved the Nordic ECOlabel SWAN, documenting its durability, health safety and environmentally sound production.

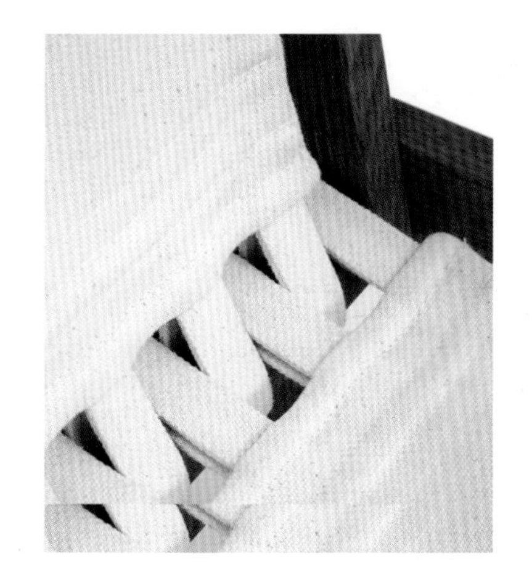

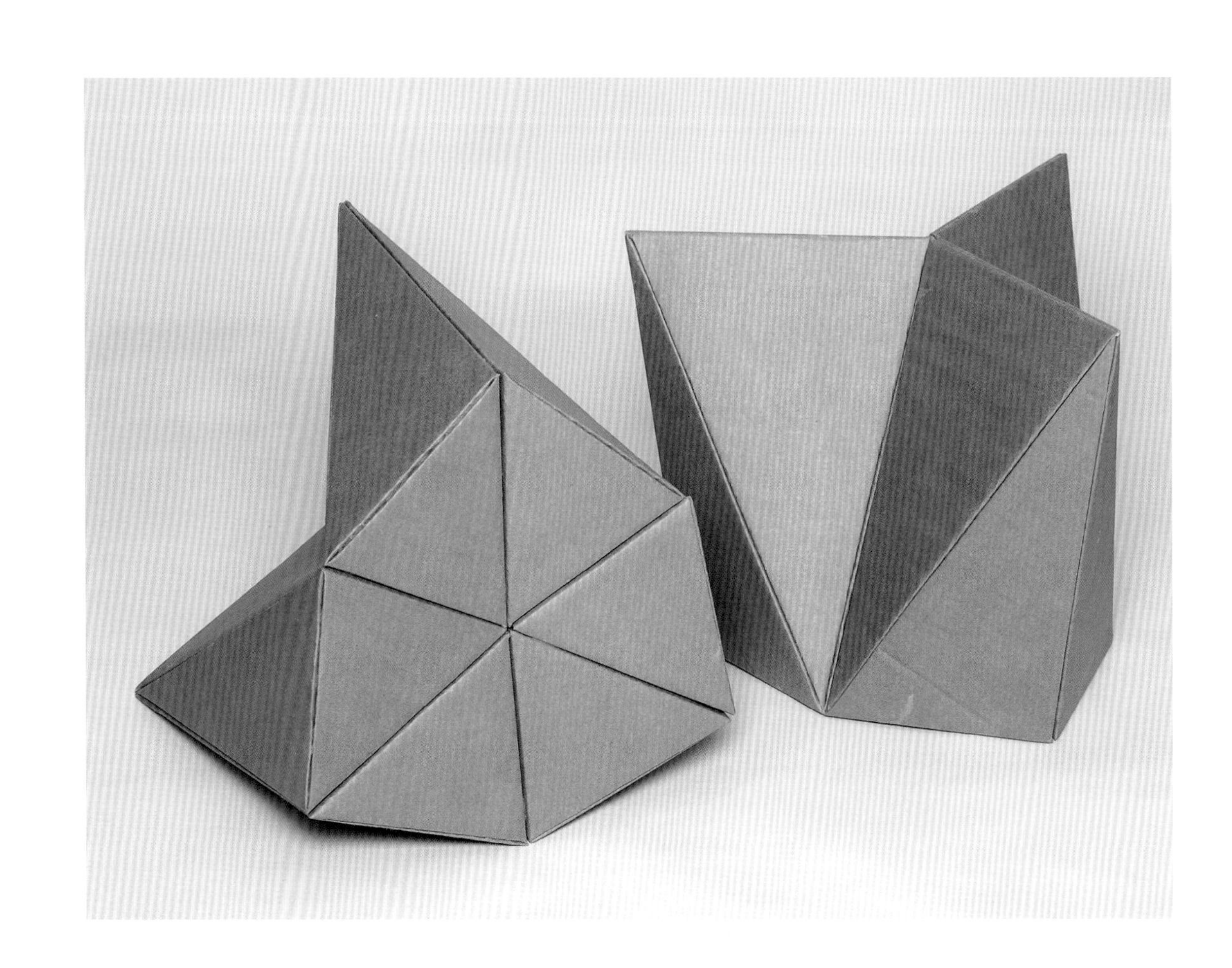

Foldschool

Foldschool

Foldschool 是一套折叠纸板组成的儿童家具，可由孩子亲手制作。网上下载图案，电脑打印。按说明操作即可组装出一个稳定的家具。

Foldschool is a collection of free cardboard furniture for kids, handmade by you. The downloadable patterns can be printed out with any printer. Follow the instructions and assemble a stable piece of furniture.

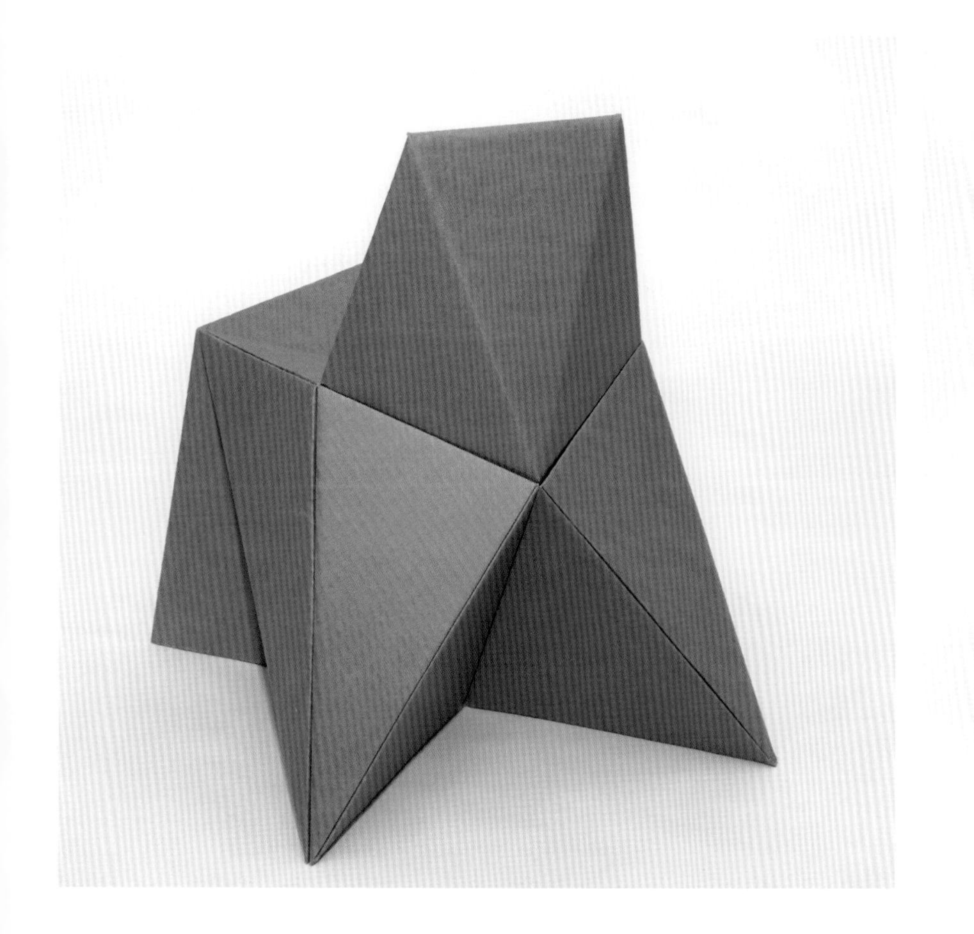

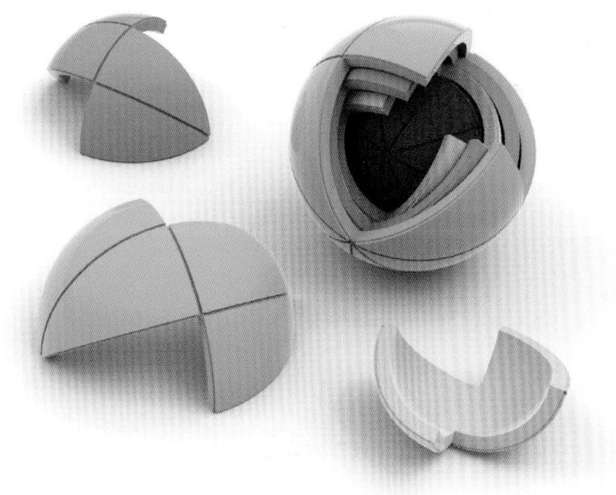

oblo™ spheres

oblo™

oblo™ 是一款为学龄前儿童设计的 3D 益智魔方。为了游戏逐级顺利进展，孩子们需要找到每个可移动模块的正确位置，从而在特定的形状空间里找到最易于插入模块的通道。

这款玩具不仅有助于训练孩子的书写绘画技巧和手眼协调能力，还能帮助他们提高智力水平和动作技能。oblo™ 魔方已申请专利保护。

Toy oblo™ spheres is designed for preschoolers. In order to obtain a successful progress from one level to the next one, one needs to find the right extracting position for each removable piece to enable its smooth passage through the fixed shape open space provided.

This is good preparation for writing and drawing, as well as developing hand-eye coordination. Toys intend is to help developing intelligence as well as fine motor skills. oblo™ spheres is patent and trademark protected product.

Balloon Bench

h220430

本设计努力营造法国电影《红气球》（1953 年）中主角的视觉体验。被吊在天花板上的气球造型长凳利用视觉错觉呈现漂浮感。

This bench was visually inspired by the feeling of floating that the main character felt in the French movie,"Le Ballon Rouge"(1953). In reality the bench is suspended from the ceiling by 4 anchors concealed by the balloon shapes. This creates the illusion of the bench being lifted by balloons.

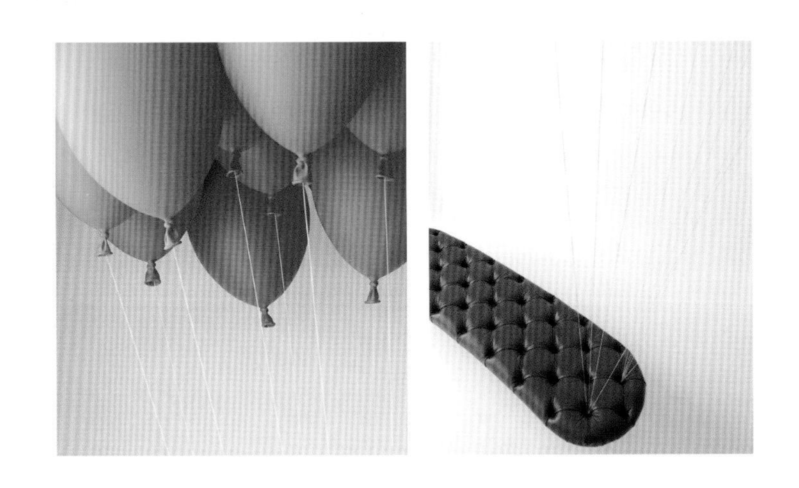

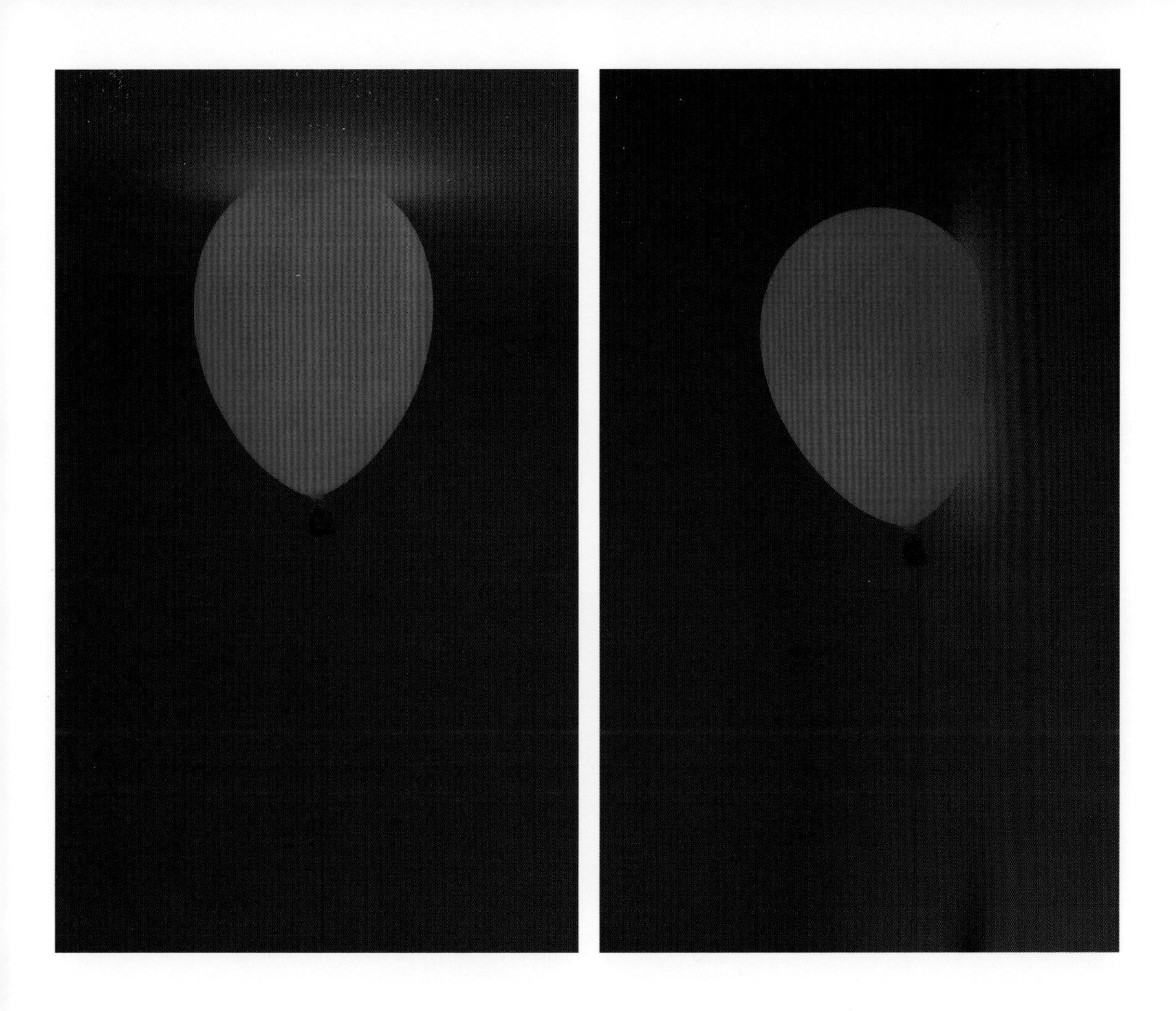

Balloon Lamp

h220430

Play

Little Red Stuga

这个胶合板制作的可折叠屏风不仅给孩子创造了个人空间，还能用作舞台布景，或搭配城堡、山峦、房屋等。Play 最初是为幼儿园孩子设计的实验性产品，满足他们在特定阶段的成长需要。

Play is a folding screen in plywood that gives personal space for the children, or it can be used as side scenes to create a scenographic atmosphere in the play. The product can associate to play among castles, mountains or houses. Play was originally part of a design examwork, developed in consideration with the needs children have in a kindergarten.

Landscape

Little Red Stuga

Landscape 地毯与三座“山峰”组合，营造出层层叠叠的山峦风景，也可以另外搭配一个或者几个厚圆椅垫，或单独作地毯使用。地毯上的手绘图中有平原、流水、铁路和道路，呈空中俯瞰风景的效果，是传统运输地毯图案的延伸。地毯还为孩子们留下了更多的想象空间，他们可以自行决定蓝色的部分是道路还是河流。

The rug Landscape completes the three mountains, and can be used together with one or several poufs, or simply as a rug by its own. Our hand drawn image becomes an aerial view of a landscape with fields, water, railways and roads and is a development of the traditional transportation rugs motifs. We have left more room for the child's own imagination by not completing everything; the child decides if the road is blue or a river.

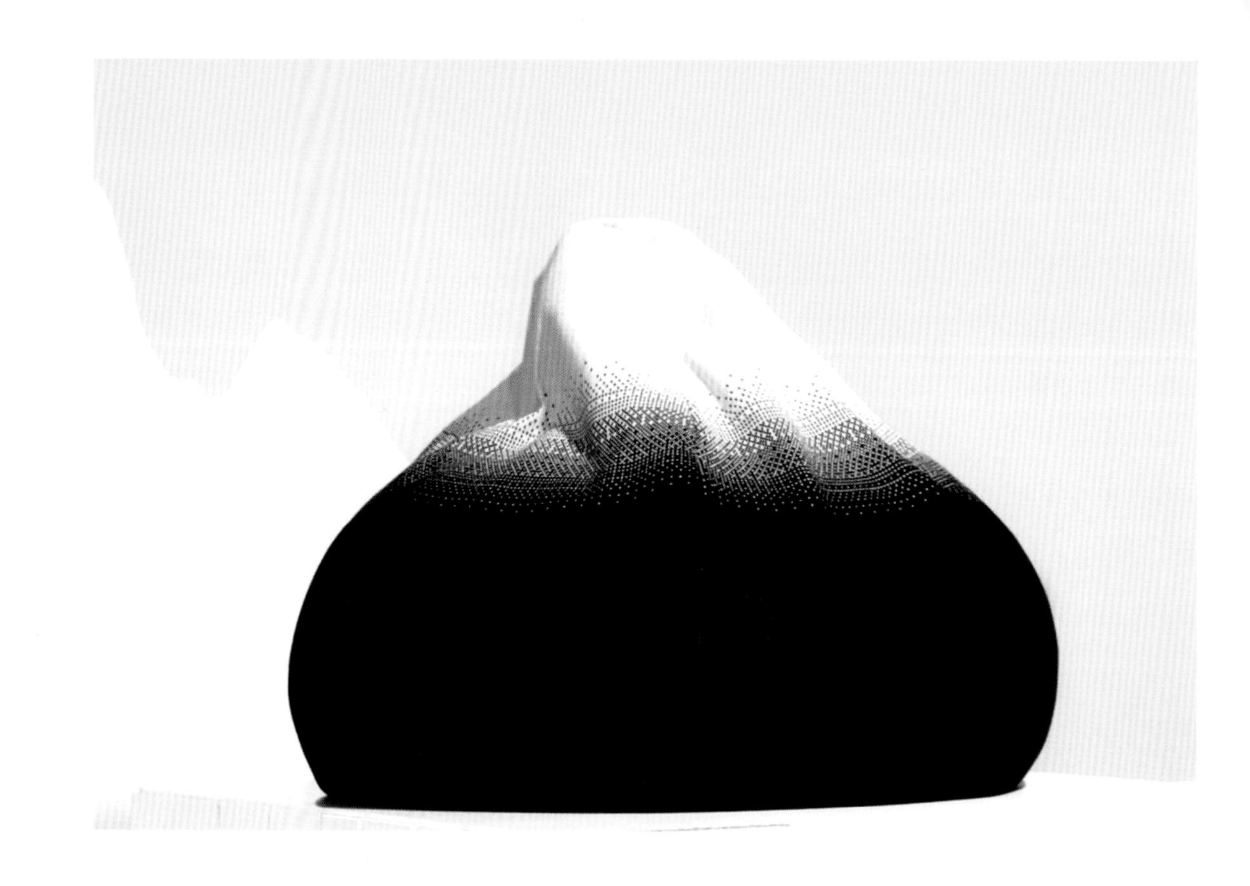

Kebnekaise

Little Red Stuga

Kebnekaise 是针织材质的厚圆椅垫，模仿瑞典北部高达 2100 米的最高山峰形状。现在每个孩子都可以在客厅里爬上瑞典最高峰了！

Kebnekaise is a knitted pouf and the highest mountain in Sweden at 2100 m and is situated in the north of Sweden. Now every child can climb it in the living room!

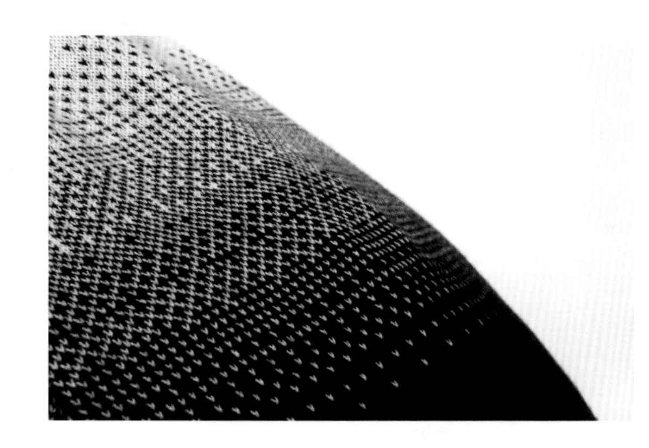

Fuji

Little Red Stuga

这款最负盛名的功能型家具在设计上的一大挑战就是要围绕豆袋坐垫概念创造出新颖有趣的产品。为了完成这个项目，我们反复探索和思考自己与高山的关系，比如，富士山、乞力马扎罗山和珠穆朗玛峰等，每个作品的颜色与布局都源自设计团队的实地考察。如果细致观察孩子们会发现许多令人惊喜的细节，比如珠穆朗玛峰上的帐篷、富士山上可爱的卡通形象或者乞力马扎罗山上的部落面具。每一座“山”都标明高度，让孩子在游戏中获得学习的乐趣。

The challenge was to create something new and interesting around the beanbag concept, one of the world's most appreciated and functional pieces of furniture ever designed. For this project we explored our own relationship and associations with mountains; namely Fuji, Kilimanjaro and Mount Everest. Each pattern and color is a result of our journey discovering them. At a close-up you find more details in the patterns, like climbers and base camps on Mount Everest, Kawaii figures on Fuji and tribal masks on Kilimanjaro. The height of each mountain is printed to add a pedagogical layer. Perfect for play and soft storytelling moments.

Dream BIG

Little Red Stuga

Dream BIG 让你梦想成真——这是一款又大又舒服的游戏毯，由柔软的高质量纺织物做成。身材高大或娇小的孩子都能在上面尽情玩耍。

Dream BIG is your dream coming true – a big comfy play mat in soft high quality fabrics! For both big and small children to hang around in.

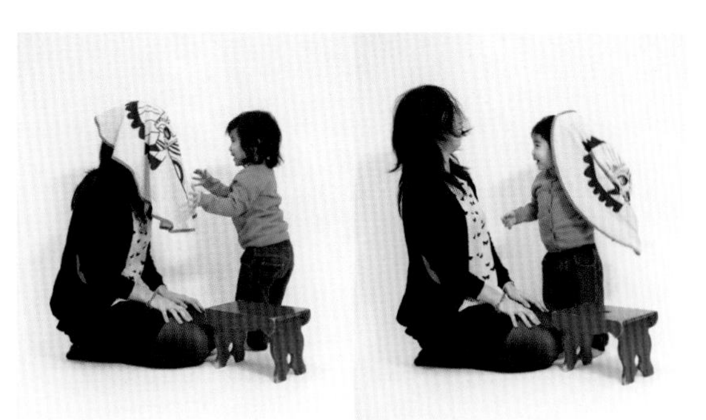

Bu!
The Blankie

Little Red Stuga

这款舒适而有灵性的披肩是每个孩子的好朋友。

百分之百纯棉披肩伴随孩子从婴儿到蹒跚学步。中间设计的两个洞洞可使披肩作为孩子们藏猫猫的道具，从而增添游戏乐趣，丰富孩子的想象力。

六到九个月大的宝贝们可以用它来玩自己的躲猫猫游戏，以训练记忆力。

一岁及以上的孩子也能从中找到躲猫猫游戏的乐趣。

三岁及以上的孩子则可以用它和同伴们一起玩角色扮演游戏。

Every child's best friend. Comfort and inspiration on the go.

This 100% cotton blankie will follow the child from infancy into the toddler years. Bu! encourages children's imagination & interaction as it doubles as a peekaboo sheet.

Babies of 6-9 months can play their own version of peekaboo, a game that stimulates the memory.

Age 1 and up — all hide and seek games are fun.

Children aged 3 and up, play a lot of character games together with other children.

Vetro Rocker

Nurseryworks

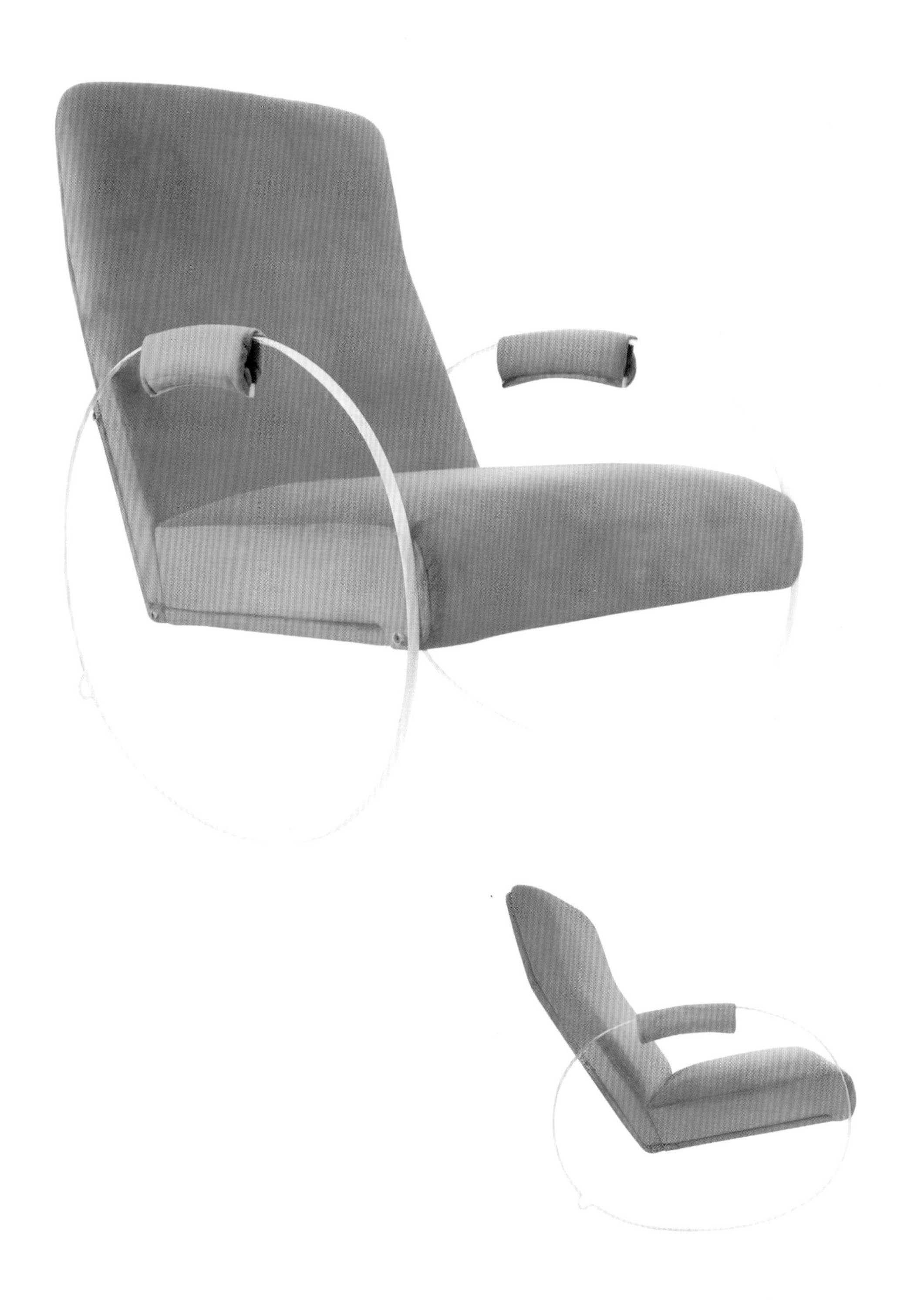

Vetro Rocker 为您带来精确的平衡感和非凡的舒适度。作为 Vetro Crib 的好伴侣，这款亚克力摇椅给人身心愉悦之感，手工制作奉献尊贵体验，经典佳作，终身享用。摇椅上的育婴生活超越时尚，与众不同。

持久耐用的亚克力侧面和底部缓冲垫确保每一次摇动都平滑安全，长毛绒座椅和软垫扶手牢靠舒适。育婴之外，Vetro 摇椅还会为个性家居锦上添花。

Precisely balanced, unbelievable comfort. A dazzling companion to the Vetro Crib, this acrylic rocker stands on its own as a one-of-a-kind triumph. The Vetro Rocker is hand-crafted to guarantee you divine rocking, on a masterpiece built to last a lifetime. Rock and nurse baby in unparalleled, ultra-modern style.

Durable acrylic sides and bottom stopper ensure smooth motion and safety. The plush seat and back are harmoniously balanced for comfort and stability, and the quality cushioning is delightfully cozy. Upholstered arm rests complete the experience. Beyond the nursery, the Vetro Rocker is a gorgeous comfort for any room in your distinguished home.

Stacking Throne

Studio Laurens van Wieringen

Stacking Throne 伴随孩子共同成长。

这套椅子由五个互相堆叠的木椅组成。

婴儿可以坐在带“T 吧”的最小号椅子里面，以防止后翻。

孩子长大后，可将“T 吧”撤出，代之以其他尺寸的椅子，分别适合不同年龄阶段儿童的成长需求。

与此同时，孩子可以按照他们自己的想象力和心情搭建一个棚子、一列火车或者一艘船，甚至可以把其中一个椅子倒挂在另一个椅子上。而其中最大的椅子能为成年人提供足够大的空间，所以这款椅子组合适用于生活的方方面面。

The Stacking Throne is a chair that grows with children.

The chair consists of 5 soft foam elements which are stacked into each other into the wooden chair.

Babies can be seated in the smallest chair element with a “T-bar” to prevent falling out.

When children are old enough the “T-bar” can be removed and from there, by taking out the 4 of 5 foam elements the chair becomes bigger as the children grow, fitting according to the size.

In the meanwhile children can play and build with the foam elements according to their imagination and mood. Build a shed, a train, a boat or they can put the foam elements upside down on top of each other and make a cabinet. When 4 of the five elements are taken out the chair is big enough to seat an adult, so the chair can be used all your life!

Inside Skate

modernconvenience

这款滑板将成为从高处滑下时最柔软的垫子。由乙烯制成的滑板是一件可骑雕塑。专门设计来为室内玩耍的冒失鬼们提供乐趣。

This is an upholstered vinyl skateboard for a soft cushioned skate down your sweeping hall. It's sculpture you can ride. Designed for the indoor daredevil and fun.

Blue

kidsonroof

用超级加固纸板制成的三维建筑，由可回收纸板制成。

A three dimensional building with super sturdy and recycled board.

Trojan Horse

koonstore

Uncensored

koonstore

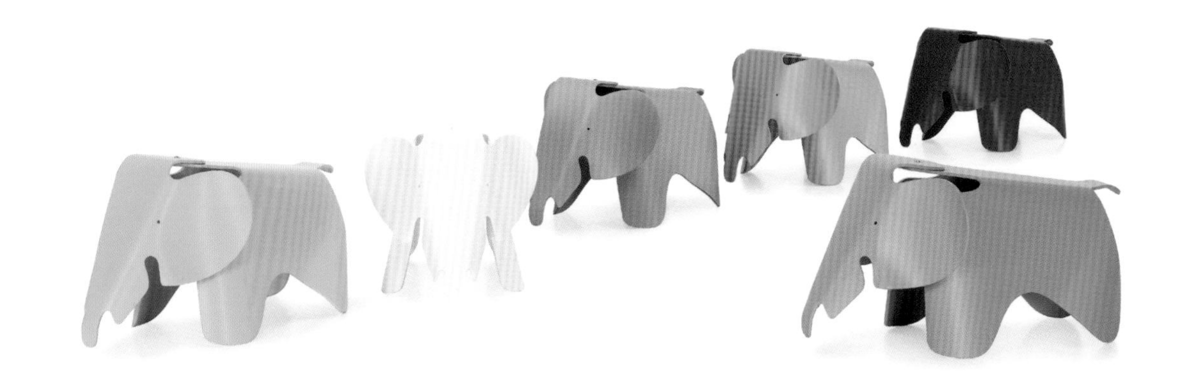

Elephant

Vitra

几乎再也没有其他动物像大象那么受欢迎了。庄严的外形和众所周知的绝佳幽默感使大象成为了孩子们每天都想拥有的伙伴——一个玩具、一个故事主角或一种神圣的动物。Charles 和 Ray Eames 两位设计师也喜欢这种动物，并在 1945 年制作了一款胶合板材质的玩具象。但此设计却从未大量投入生产。现在，这款为孩子而设计的玩具象的塑料版本首次面向市场投产。

Almost no other animal enjoys such popularity as the elephant. Admired for its majestic size and loved for its proverbial good-humour, it is part of our everyday experience as a child's cuddly toy, a storybook character and a majestic creature. Charles and Ray Eames also succumbed to their charms and in 1945 designed a toy elephant made of plywood. However, it never made it into mass production. The Eames elephant is now available for the first time in a plastic version for those it was originally intended for: children.

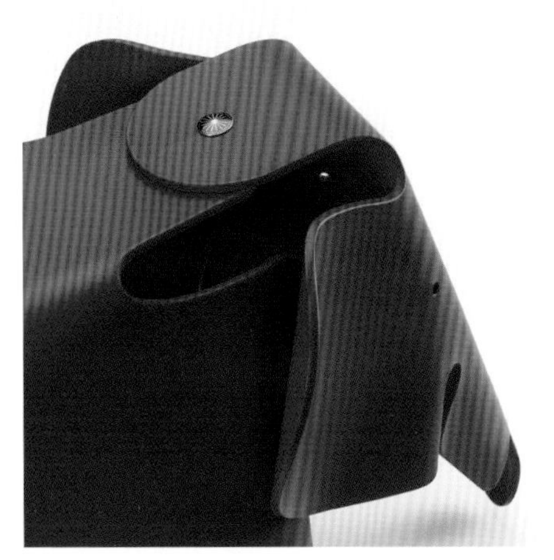

Panton Junior

Vitra

Panton 椅子一直以来都是孩子们的最爱。他们不仅喜欢它明亮欢快的色彩、光滑的触感和弯曲的形状，也着迷于把玩它和坐在上面的巨大乐趣。它不仅可以作为一个椅子，也可以作为巨大的洞穴和躲藏的地方。本产品推出后，Verner Panton 打算与 Vitra 公司联合推出一款适合孩子尺寸的椅子。

The Panton Chair has always been a favourite of children. They not only like its bright, cheerful colours and smooth curves, but the fact that it is as much fun to play with as it is to sit on. Soon after its introduction, Verner Panton began to consider the idea of producing a child-size version of the chair together with Vitra.

Circle

COLLECT Furniture

本款家具专为一至五岁儿童设计，由一个圆凳和中心桌组成。孩子们可以爬到里面坐着画画，也可以在周围和上下玩耍。

迄今为止，产品仅在丹麦生产，由中密度纤维板制成，能够与任何装修风格及各种颜色的壁纸搭配。

Circle is designed for children 1-5 years old. The furniture is composed of a round bench with a central table. Children climb into the furniture to sit and draw, but just as much to play inside, on top and around it.

Until now, Circle is only produced in Denmark upon request and is delivered in untreated MDF-furniture plates to paint, decorate or wallpaper in any choice and color.

Totem City

Kidsonroof

Totem City 是用超级加固纸板制成的三维建筑。它的 130 个部件双面均刻有宝剑、巨龙、神秘符号和纹理。此包装中的部件可用于建造四种模型：教堂、隧道、机场和“牦牛”（不能同时造出四种模型）。包装盒中配有说明手册。最重要的是，孩子可以充分发挥自己的想象力和创造力！

适用于六岁以上的孩子，有 130 个部件，由可回收纸板制成。所有 Totem 系列的部件都互相兼容。

Totem City is three dimensional building with super sturdy cardboard. 130 pieces reversibly printed with swords, dragons, secret symbols and textures. Build all 4 models — a church, a vessel, an airplane and a "yak" — with the pieces from 1 box! (Not all at the same time!). A manual is included. But most of all...Feel free to make any of your own fantasy creations!

For kids 6+. 130 pieces. Made of recycled cardboard. All Totem sets are compatible.

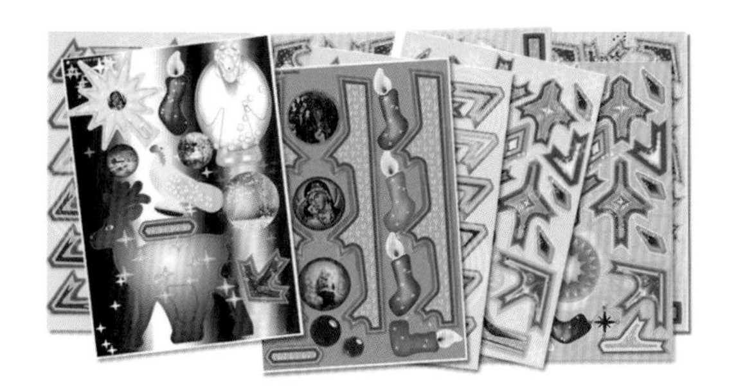

Xmas

Kidsonroof

由超级加固可回收纸板制成的圣诞装饰。

Christmas decorations made of super sturdy and recycled cardboards.

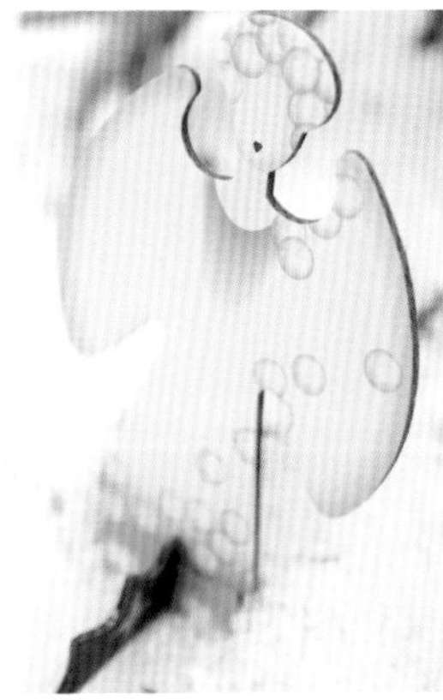

Figurines

Kidsonroof

由超级加固可回收纸板制成的小摆件。

Figurines made of super sturdy and recycled cardboards.

Dragonfly

Kidsonroof

由超级加固可回收纸板制成的蜻蜓。

A dragonfly made of super sturdy and recycled cardboards.

Sail

Kidsonroof

由超级加固可回收纸板制成的帆船。

A sailing boat made of super sturdy and recycled cardboards.

Leander Stool Set

Leander

众所周知，小孩子都很好动，即使坐着也不会安安静静。因此，我们开发的这款全新凳子组合以人体工程学原理为基础，造型优美和谐。

孩子坐在凳子上时，身体压下椅子前侧；站起来时，椅子自动恢复到原来状态。这款凳子的动力学设计使孩子们早早就学会如何保持自然舒服的坐姿，为他们发挥创造力提供了理想条件。

不用时，两个椅子可以推到桌子下面，以节省空间。这款凳子套装有纯白色、石灰色和樱桃木花纹三种选择。

Everyone knows that children move around, even when they are sitting still. Therefore, we developed this new, ergonomic stool set that is simply beautiful and harmonic.

As children sit on the stools, their weight presses the front edge of the seat down; when they stand up again the seat automatically returns to its original position. The stools' dynamic function gives children a naturally good sitting position in a young age and the ideal conditions for exploring their creativity.

When not in use, both stools can be pushed under the table so that they do not take up more space than the table itself. The stool set comes in white, whitewash and walnut.

Colorstones

smarin

Colorstones 是一个十年前就推出的 Livingstones 系列新版本。公司的合作伙伴 Maison Kvadrat 公司已经同意我们生产 Divina 和 Hallingdal 这一线产品的彩色版本。

简单的有机造型涂上完美调色涂料，将会营造出全新的风景。

这些大大的五彩石头看上去既动感十足，又宁静祥和，一定会让孩子的游乐空间充满情趣。

Colorstones are a new version of the Livingstones conceived almost ten years ago. Our collaboration with the Maison Kvadrat has allowed me to produce a colored version of the famous pebble cushions in the Divina and Hallingdal lines.

Their simple and organic shapes, interpreted in a magnificent color palette, will paint new landscapes!

Each space inhabited by these massive, colorful, dynamic yet peaceful shapes is going to acquire a new dimension: emotion.

储 物 的 设 计

Products for Storing

儿童自用储物用具的主要目的是培养孩子学习整理自己的用品，因此趣味性、易识别性、简洁性、安全性便成了设计的要点。组合方式、稳固程度、形状、色彩、高度、材质、放置位置、归类方式和功能都需要根据儿童的特点认真考量。一般来讲小一点的孩子储物用具较简单，如鞋架、衣架、垃圾桶、玩具箱等就足以让他们摆弄一阵子了；大一些的孩子可以给他们提供衣橱、书柜等家具，让他们学习打理自己的生活，使生活更有条理。

不论设计何物，儿童产品的趣味性是很重要的原则。趣味性决定儿童是否愿意使用。趣味性的重要特征是故事性、拟人性、夸张性、梦幻性、醒目性或带有一定的卡通性。

本章为读者提供了一些优秀的设计产品，希望能够启发读者的思路和使用的选择。

The storage equipments for children are designed to cultivate the ability of dealing with their business and thus featured with being interesting, legible, simple, and safe. Besides this, other elements as the combination mode, stability, shape, color, height, material, position, category and function are also needed to taken into serious consideration. Generally speaking, younger children are pleased to use simple storage appliances like shoes rack, clothes rack, trash can and toy box, while the older ones are happy with wardrobe and bookshelf, etc., to take care of their own articles for daily use.

All children products are expected to be interesting, playful and attractive, supported by stories, personification, exaggeration, dreams, or cartoons.

This chapter aims to enlighten you with some outstanding design works.

MY STORAGE

Magis

设计师试图创造一种智慧、简单且玩性十足的工业生产系统，以最少组件搭配最多使用场合。

MY STORAGE 适合各种合作环境，老少皆宜。

用户可以用它制作抽屉、开放性储存空间或小型书桌；也可以根据自己的喜好和需求调整选择两种不同高度，使用不同组件。

The designer wanted to make a clever simple and playful industrial produced system, that could be used in many different ways with a minimum of components.

The designer think MY STORAGE would fit in domestic and contract environments and could be good for children and adults.

You can create chest of drawers with it, or open storage and even small desks. Through the height of the poles you can choose two heights and apply the different components to your own taste and request.

Boom Cabinet

Straightline Design

Boom Cabinet集趣味性与功能性于一体，每一个“爆炸”抽屉都附着在墙壁上，打开后即可存放物品。

Fun and functional is the idea behind the Boom Cabinet. Each “exploding” drawer attaches to the wall, and opens up for storage.

Hollow Chair

Straightline Design

这件作品由超过 600 片环环相扣的胶合板组成，可以把自己喜欢的任何东西放进去创造一个崭新的空间。

This piece was made with over 600 interlocking pieces of plywood. It is built so that people can put in anything they like, and integrate the piece into their space.

Beaver

Straightline Design

Beaver 橱柜是一款为人津津乐道的作品，它做工精细复杂，由东部枫木、枫木贴面、两部分泡沫和玻璃纤维构成。

The Beaver cabinet is very a labour intensive piece. It is made from eastern maple, maple veneer, two part foam and fibre glass. The final result is a piece of furniture that will always start a conversation!

Raymond

Straightline Design

这款富有想象力的作品实际上是一个“坐”在地上的衣柜，外观随和，人见人爱。

This imaginative piece actually sits on the floor. It is an easy going cabinet that anyone will talk about when they see it.

Carrie the Crayon

Straightline Design

本设计为 Crayola 始建，由一个巨大的蜡笔盒抽屉和一个连接面板构成。

This design was originally built for Crayola. It is a giant crayon box with working drawers, and an access panel.

Carrot Cabinet

Straightline Design

这款外形古怪的木骨架柜健康实用，叶子在泡沫造型基础上采用玻璃纤维雕刻而成。

This was a unique piece made from a wood skeleton cabinet. We blow on foam, and then sculpt using fibre glass. It is a whimsical, healthy cabinet!

Accordion Cabinet

Straightline Design

一款由西方枫木和枫木贴面制成的手风琴橱柜，虽不能演奏乐曲，却能为孩子带来无限欢乐。

Although it doesn't play music, it is sure to make you smile! The Accordion cabinet is made from Western maple, and maple veneer.

Sahara Chest of Drawers™

argington

这款经典育婴梳妆台是存储尿布、湿巾、面霜和衣服的理想选择。Anywhere Changing Tray™ 特别强调舒适度和安全性，保证 Sahara Chest of Drawers™ 可做可变桌使用。当孩子逐渐长大不再适合 Anywhere Changing Tray ™ 时，可将其重新组装成梳妆台，供学步阶段孩子使用至成年。Sahara Chest of Drawers™ 顶部表面光滑，用户可根据自己的喜好在上面摆放台灯、相框或其他装饰物件。

The Sahara Chest of Drawers™ is a classic dresser for your baby's nursery. It is ideal for storing diapers, wipes, creams and clothes. Our Anywhere Changing Tray™ fits snuggly and securely on top, allowing the Sahara Chest of Drawers™ to be used as a changing table. Once your child has out grown the Anywhere Changing Tray™, it converts conveniently into a dresser that will be used from toddler to adult years. The Sahara Chest of Drawers™ has a flat surface top giving the ability to style the dresser by adding a lamp, picture frames or decorative items. The Sahara Chest of Drawers™ can fit anywhere in a house.

Training Dresser

peterbristol

手工制作，东华盛顿大山视野细木工家具公司精心设计包装。

橱柜由 1.9 厘米厚 ULDF 制成，刷改良光油漆，抽屉由 1.27 厘米层压槭树胶合板制成，催化固化漆接合。所有橱柜与抽屉均采用数控切割、打孔、开槽。同时配备螺丝、大头钉、胶水以及来自太平洋西北部的新鲜空气。

Well considered and well made. The dresser is hand crafted and packaged with care in Eastern Washington by the crew Mountain View Cabinetry.

The cabinet is made from 3/4" ULDF and finished with conversion varnish. The drawers are 9 ply 1/2" maple plywood, dovetailed and finished with clear catalyzed lacquer. All cabinet and drawer components are cut, drilled and dadoed on a CNC table router. Assembled with a combination of screws, pins, staples, glue, and Pacific Northwest fresh air.

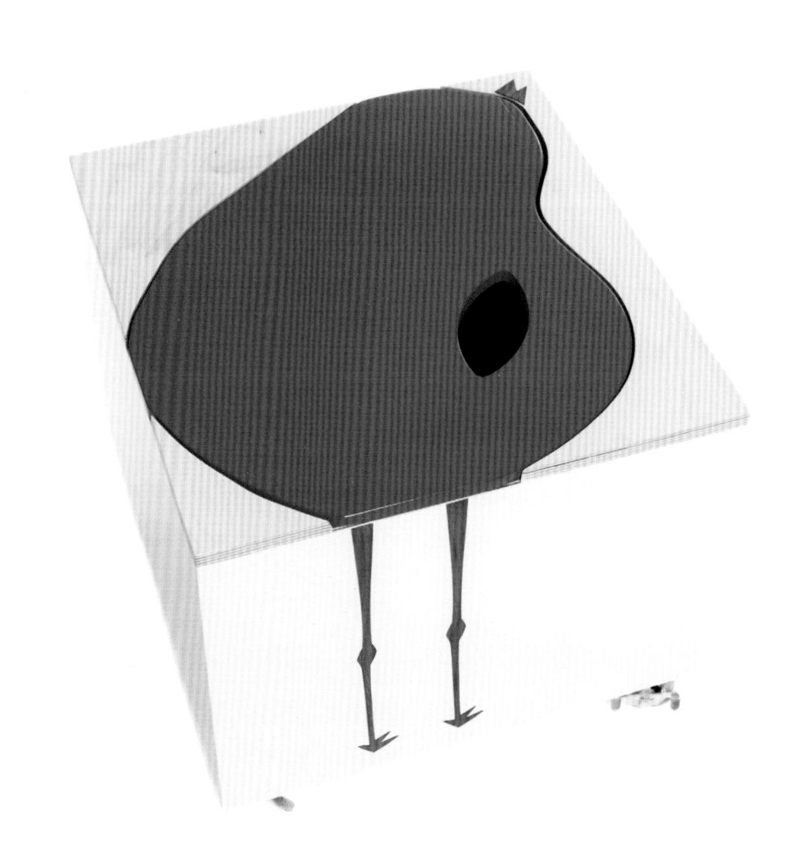

The Bertie Box

modmomfurniture

The Bertie Box是短期存储的理想选择。它贴近地面，轮子方便移动，顶端的拼图设计使安放玩具的过程更加充满乐趣。

由环保的桦木胶合板制成，柜子内部为天然桦木，翅膀状的开口便于轻松提起盖子。整个盒子由低VOC（挥发性有机物）的水性聚氨酯涂层覆盖，以提供光泽保护。阿米什工艺，美国制造。

The Bertie Box is a perfect storage option for the short set — it's low to the ground, sits on casters so it moves easily about the house, and the puzzle piece top makes putting toys away more fun!

Made from eco-friendly birch plywood. Natural birch interior of box. The wing cut-out acts as the hand-held opening for lifting the lid. Entire box is coated with low VOC, water-based polyurethane for sheen and protection. Handmade in the USA — Amish-made.

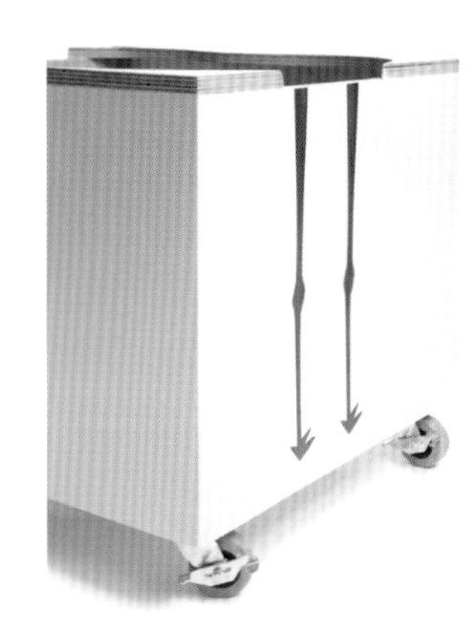

Owyn Toy Box

Mod Mom Furniture

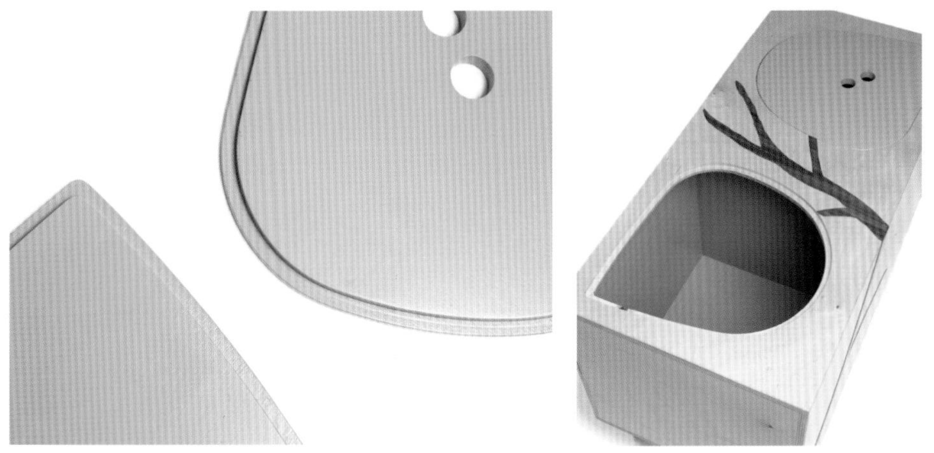

Owyn Toy Box 是一个现代感十足的有机存储箱，可容纳大量玩具和家庭用品，有两个易开式盖子，盖子打开盖上像是在拼拼图。

产品由环保桦木胶合板制成，环保漆涂刷内部。整个盒子由低 VOC（挥发性有机物）的水性聚氨酯涂层覆盖，以提供光泽保护。阿米什工艺，美国制造。

The Owyn Toy Box is a modern organic-style storage piece designed to house lots of toys and/or household stuff. It has two easy lift-off lids that lift off and place back like a puzzle. The toy box is separated into two convenient compartments under individual leaf lids.

Made from eco-friendly birch plywood. Green painted interior. Entire box is coated with low VOC, water-based polyurethane for sheen and protection. Handmade in the USA using Amish carpentry techniques.

cational and engaging furniture for children's bedroom

ELENA NUNZIATA

每样物品都在讲述一个引人入胜的故事。

Paddy 收纳箱长着一双可旋转的按钮“眼睛”，一直专注地为主人讲着故事，而那不时瞟着所有衣钩的眼神使他的故事更加生动。

矮衣帽架便于孩子通过调节挂棒选择适合自己的使用角度。

查理是出了名的“脏衣服消化桶”，莱卡麻袋肚子总是撑得饱饱的。孩子们喜欢给他塞饱的感觉，也学会了该何时和父母一起洗干净这些衣服。

梅尔文这个神奇的床头柜是一件很实用的家具，上面的床头灯可以满足孩子的阅读要求，灯上的“保护神”可以整晚守护着孩子。

所有作品均以榉树为原材料，手工制作，辅以数控和激光雕刻技术。

Behind each object lies a story that engages the child in its use.

The paddy collection has eyes as knobs that can be spun around. The endless narrative of them staring and following the user is enhanced by the ability to slide from side to side for all of the hooks of the coat rail.

The short coat stand allows the user to move the rods through the central pole and play adjusting their position as preferred.

Charlie is best known as the "dirty clothes eater". A lycra sack allows clothes to cumulate until its belly is full. Children can enjoy to stuff him and understand when it is time to load the washing machine with their parents.

Melvin the magic bedside table becomes a useful and desirable piece of furniture. It embodies the functionality of a bedside lamp, good for reading stories, and the metaphoric figure of a night guardian that watches over the child.

All the pieces are hand-made in beech with the aid of CNC and laser engrave technology.

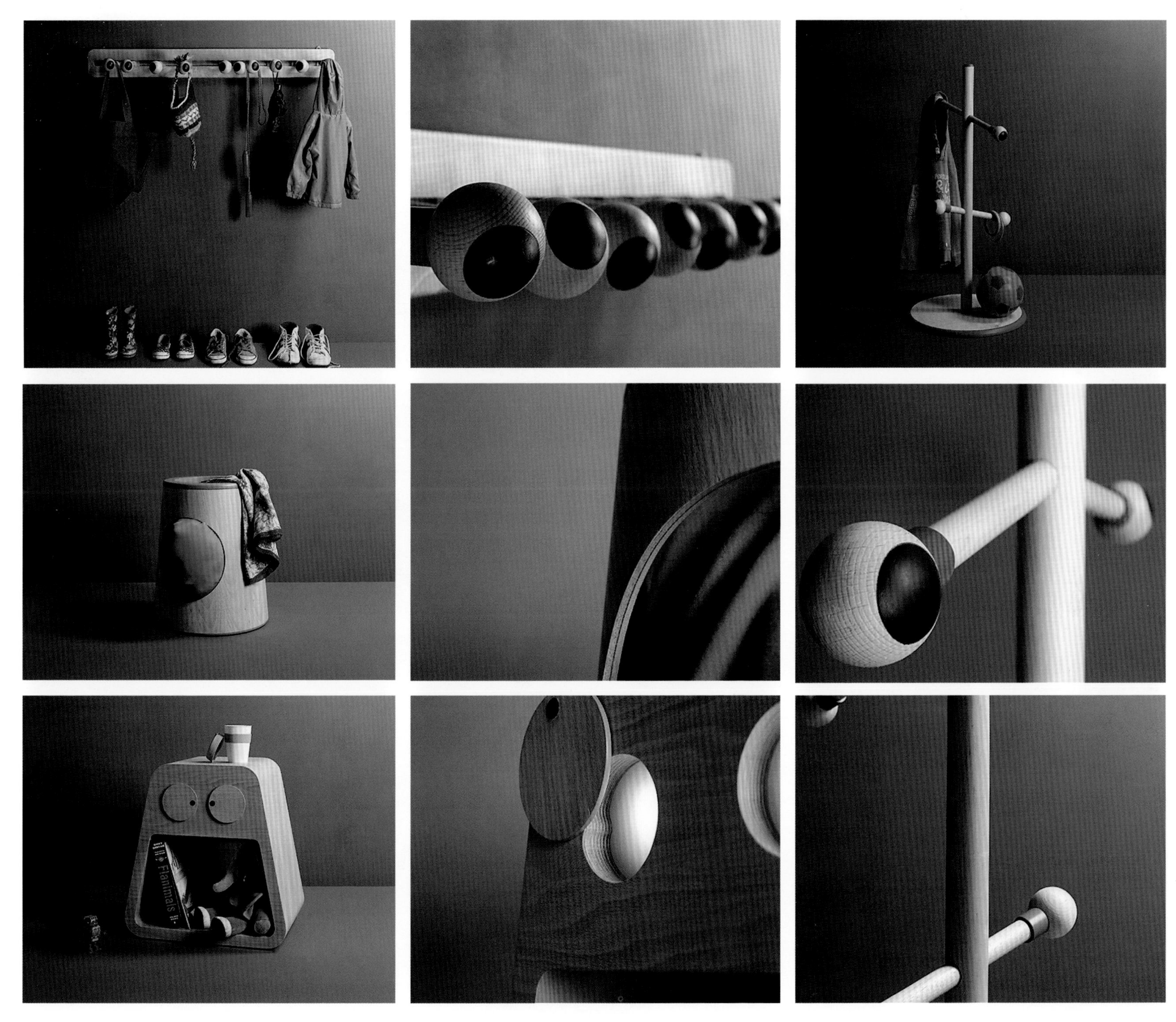
Flanimals

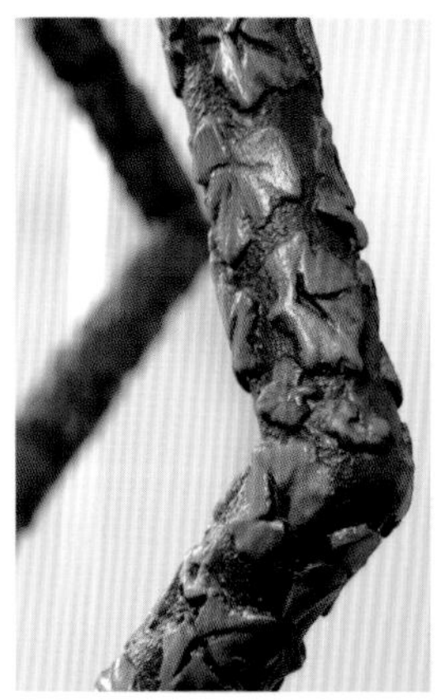

YVR

Straightline Design

这套自定义安装家具建于 2009 年，包括迷宫、壁炉电视、拖轮、惠斯勒墙、码头游乐和灯塔。这些精彩的作品为成千上万的儿童带去欢乐。

These custom installations were built in 2009. They include a labyrinth, fireplace television, tugboat, Whistler wall, dockside play area, and lighthouse. These wonderful pieces bring smiles to thousands of children!

Bug Street

Straightline Design

这个漂亮的存储单元中的每个建筑都自成一体，方便组装拆卸，孩子们一定愿意把自己的玩具放到里面。“建筑”底层有可存放各种物品的小储存间，“建筑标牌”可按客户要求制作，如以孩子或宠物的名字命名。

Each building in this beautiful storage unit can come apart for easy assembly/disassembly. Kids will want to put their toys away in this unique piece. There are storage cubbies underneath for a variety of items. The signs on the buildings are custom, and can be made with any child's or pet's name.

Tree Bookcase

Nurseryworks

这款树形书架的设计灵感源于对阅读的热爱。洋溢着大自然气息的书架一定会激发孩子的想象力，使他们愉快地享受从“树杈”上选书的乐趣。

树形书架采用零级中密度纤维板制成，经久耐用，可安全固定在墙上。书架纯手工组装，砂纸磨光，无毒漆喷涂，确保产品光滑完美，充满活力，可与任何育婴家饰搭配使用。

True love for reading takes root with the Tree Bookcase. Nature-imbued shelves inspire imagination and fancy; you and your children will love selecting from Tree Bookcase's cheery branches.

Safely arrange over 100 books on its stable shelves. The Tree Bookcase is crafted from durable zero-grade MDF and attaches securely to the wall. This complex piece is sanded and assembled by hand to ensure you receive a perfect, smooth form. Finished in vibrant, non-toxic lacquers, Tree Bookcase coordinates with all Nurseryworks pieces.

Vetro Changer

Nurseryworks

客户可按照需要定制与 Vetro Changer 搭配使用的现代化可变桌（或储存柜），完美配合 Vetro 婴儿床，转换功能与模块设计，便于拆卸的亚克力托盘可转换为装常用品的车厢。储存柜样式可选择三个抽屉加带亚克力书架的橱柜或六个抽屉。

Customize your modern changing table/storage combo with the Vetro Changer. The perfect, subtle complement to Vetro Crib, this ultra-contemporary changer features acrylic peek-a-boo cabinet door cut-outs and a removable acrylic changing tray. Available storage options are three drawers plus cabinet with acrylic shelf, or six drawers.

A Wardrobe

COLLECT Furniture

两扇柜门共同围绕一个中心轴敞开，衣柜分为两个隔间，一侧为搁架，另一侧为悬吊衣架。

由弗雷德里克·柯莱特设计。

Two doors of wardrobe open around a central shaft. The chest is divided into two compartments, storage rack and suspension hanger. This is designed by Frederick Collette.

Green Chest of Drawers

Atelier Charivari

20 世纪 50 年代的青瓷绿柜体，镀金把手。

1950's chest of drawers in a Celadon Green, handles are kept gold

烘 托 环 境 的 装 饰 品

Products for Decorating

这里指的环境因素主要是视觉因素。

儿童生活环境是引导情绪的重要因素。儿童因年龄不同对环境的要求差异很大，不是所有年龄段的儿童都需要梦幻般的色彩亮丽的鲜艳的强刺激的动感的环境气氛。为了保证新生儿的睡眠较静的环境气氛比较适宜，强烈的视觉和动感刺激是无益的，对有听故事要求的儿童来说，尤其在二三岁之后，慢慢随着故事的复杂相应地调整环境气氛是非常有益的。不同的房间应按其功能布置，在布置的时候最好让儿童一起参与，这样可以避免按大人的喜好来代替儿童喜好。大人们应该让孩子告诉你他们需要什么。由此，设计师应该考虑儿童的参与性，墙面、地面、挂件、摆饰等都应该考虑他们的参与因素：易贴、易放置、易更换。

Environment factors in this chapter refer to visual factors.

Living environment can easily affect the children's mood and thus children of different ages ought to be exposed to various environment. For instance, bright and vibrant colors do not suit newborns in that they need quiet and mild atmosphere for sound sleep. But to the older ones who love to listen to stories, it is wise to change the environment in response to the story plots. At the same time, children had better participate in home decoration to know what they want and what they can create. For this reason, more participatory ideas should come to designers when they devise the wall, the floor, the pendants and other decorations.

Flensted Mobiles

flensted-mobiles

Flensted Mobiles 已有近 60 年为各年龄段消费者设计风铃的历史。风铃会为房间带来活力与动感，并令人身心平和，不但适合老年人与孩童长时间注视玩赏，也会使商务人士讲电话时变得更加沉静。因此，医院与诊所经常悬挂风铃，以缓解老人与孩子的紧张情绪，便于解压。

风铃虽小，尽观大千世界。我们竭力为全世界消费者制作尺寸风格各异的风铃。最近为纽约古根海姆博物馆制作的一款风铃就堪比视觉诗歌，谱出动静协奏曲。

从装饰经理办公室的抽象造型到取悦儿童的蝴蝶、瓢虫和大象等卡通形状，Flensted 风铃致力于创作“运动中的艺术”，风铃的任何轻微晃动都会让封闭房间的气氛瞬间活跃，它们或抓人眼球，或振奋人心，或平静灵魂……

Flensted Mobiles have made mobiles for all ages for nearly 60 years. A mobile brings energy and movement to a room. Little children as well as old people can gaze for hours at a mobile. Also business people can be calmed to look at a mobile while talking on the phone. A mobile brings harmony and calmness to your mind and soul. A mobile is also said to be a pain killer — again children and old people get relaxed and feel better when gazing at a mobile. They are often used in hospitals and clinics.

A mobile is a tiny interconnected and interrelated universe. We make mobiles in all sizes, materials and styles for customers all over the world. Lately a mobile made for Guggenheim Museum New York. The artistic output — comparable to visual poetry creates movement and tranquility at one and the same time.

From abstract shapes, appropriate to an executive's office to whimsical butterflies, ladybirds and elephants designed to delight children, Flensted Mobiles are art in motion. They are so perfectly balanced that even the slightest movement in a closed room sets the elements astir — drawing the eye, delighting the spirit, and calming the soul.

Wallpaper Silhouettes by InkeHeiland (1975, Kiel, Germany)

inke

InkeHeiland 墙壁剪纸纯手工制作，主要材料为稀有仿旧布料和设计师墙纸，每种款式设计包括一个 DIY 套件、刷子、说明书和环保壁纸胶。

墙壁剪纸式样繁多，可按客户要求定做，保证款款独特。

剪纸形状主要为野生动物、豢养动物、树木、灯具等。The Town Musicians of Bremen-series 故事墙纸附赠一本由 Inke 制作插图、Grimm 兄弟创作的绘本（ISBN 978-94-6150-002-1，阿姆斯特丹 David & Goliat 公司出版，有英文版和德文版）。

The wallpaper silhouettes by InkeHeiland are hand made from rare vintage and designer wallpapers. Each design comes as a DIY-kit, complete with a brush, instructions and some eco-friendly wallpaper paste.

Because of the many possible variations, the silhouettes are mostly made to order. This makes each piece truly unique.

Designs include wildlife and domestic animals, trees and lamps. The Town Musicians of Bremen-series is accompanied by a colorful children's book, based on the tale by the Grimm Brothers, illustrated by Inke(ISBN 978-94-6150-002-1 David & Goliat Publishing Amsterdam — also available in English and German).

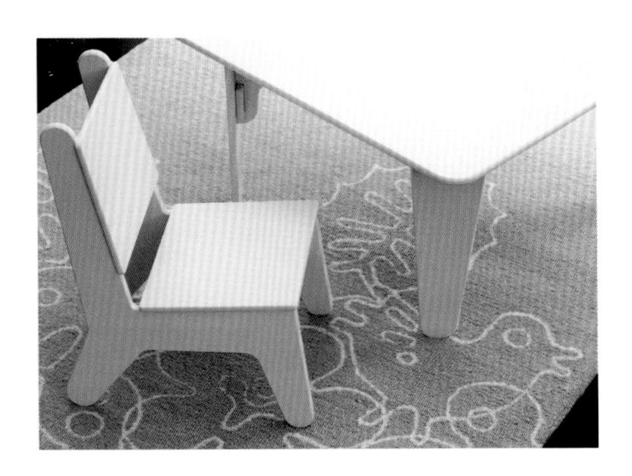

notNeutral

notNeutral

所有的 notNeutral 产品均为纯羊毛制品，着无毒染剂，可循环棉布底衬。地毯通过 GoodWeave 认证，保证在加工过程中没有童工参与。GoodWeave 一直致力于帮助童工，并为那些从印度、尼泊尔、巴基斯坦解救出来的童工提供教育和康复援助。

All notNeutral rugs are 100% wool, use non-toxic dyes, have recycled cotton backing. The rugs are GoodWeave Certified, assuring that no children were employed by the facility responsible for making the labeled rug. Good Weave products are committed to the deterrence of child labor, and the education and rehabilitation of rescued child workers in India, Nepal and Pakistan.

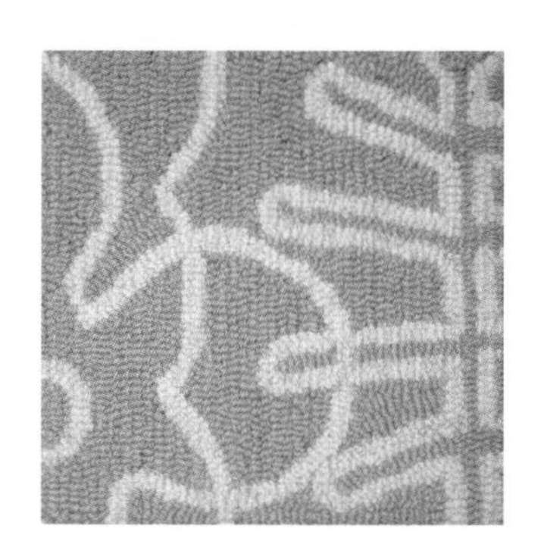
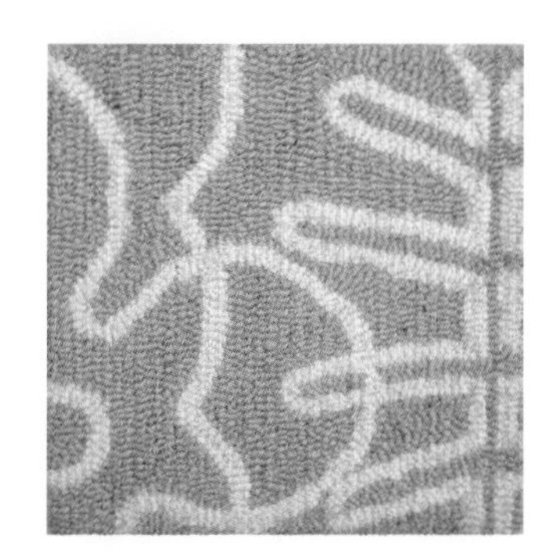
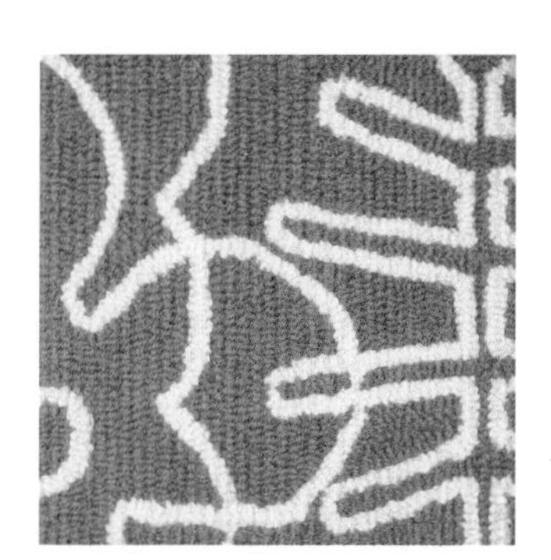

Animal Cushions

Scoops Design/ Sophie Cooper

电脑打印图案，手工缝制填充垫，每款均双面缝制，图形相同，色彩相异。

Digitally printed, hand sewn and stuffed cushions. All Designs are double sided with reverse colours on the back.

Shatterproof Mirrors

Scoops Design/ Sophie Cooper

激光切割银镜玻璃而成，每款设计为一家三口的造型。

Shatterproof mirrors, laser cut from silver mirror Perspex. Each design comes in a family of three.

看 护 的 延 伸 品

Products for Nursing

儿童看护用品实质上是大人肢体或功能的延伸。本章只列举了婴儿车。婴儿车设计的因素很复杂，这里仅就与儿童和看护者相关的问题略说一二。童车的设计不但要考虑不同年龄儿童的各种乘坐方式，不同姿势所带来的安全性、舒适度问题以及儿童睡觉和玩耍时的要求，还要考虑便于携带、稳固、推行、存放、大人与孩子的视觉关系。设计师应考虑功能、尺寸、材料、颜色、结构、折叠方式、乘坐方式、重量、推行速度以及款式均需要与儿童年龄相匹配。

The products for child nursing are actually the extension of adults' limb function. This chapter illustrates the designing of strollers which requires special attention to the way of baby's sitting, safety, comfort, sleeping and playing. In addition, the strollers are expected to be easy to carry, to push, to store, and to offer parents a better view. Accordingly, the function, size, material, color, structure, way of folding and sitting, weight, speed of pushing and style of the strollers must be designed in accordance with the age of the baby concerned.

Bugaboo Cameleon

Bugaboo

Bugaboo Cameleon the all-in-one-and-only.

This versatile, modular and multi-terrain infant-to-toddler stroller is celebrated by parents worldwide for its innovative design, the use of premium materials and the endless colour as well as accessory combinations. The Bugaboo Cameleon takes you everywhere you want to go: city to country, Winter to Summer, on the beach or in the snow.

集各种功能于一身的独特产品。

多功能、组装式、多地形折叠婴儿车以其创新设计、优势材料、诸多色彩选择以及各种组合配件而受到世界各地父母的青睐。Bugaboo Cameleon 可以在任何季节带孩子到任何想去的地方——乡村、城市、寒冬、盛夏、海滩、雪地……

Bugaboo Bee

Bugaboo

Bugaboo Bee the compact yet complete stroller for parents who live life on the fly.

The Bugaboo Bee is the answer for parents who want a compact, nimble, one-piece foldable stroller that keeps up with the pace of modern parenting. Easy to weave through crowded subways and fit into car trunks, the Bugaboo Bee is a compact and maneuverable stroller for life on the fly. Its expandable design accommodates a child from birth to approximately the age of four (0-17 kg) and has many of the features and functionality expected from Bugaboo.

简洁而完备的 Bugaboo Bee 是忙碌的父母的最佳选择。

Bugaboo Bee 迎合现代育婴理念，设计出结实、灵活、小巧、可折叠婴儿车，能在拥挤地铁中轻松折叠或放入车厢以适应忙碌的生活节奏。婴儿车的各种延伸功能适用于新生儿到四岁左右孩童（0—17 千克），无论是功能性和外观设计都达到用户的期待值。

Leander High-chair

Leander

Leander 高脚椅舒适有弹性；可充分满足孩子的成长需求；根据需要自动调整安全柄的角度，无需另外借助工具；Leander 高脚椅亦是成年人的理想座椅。

好动是孩子的天性，想让他们在餐桌前安静片刻也绝非易事。我的初衷是想为孩子们制做一把宽敞舒适的椅子，后来发现型木的材料既环保又有柔韧性，用它做成的椅子弹性恰到好处，孩子们坐上去安静又舒适，有机造型给人柔和之感，与 Leander 的其他家具形成审美上的一致性。靠背与安全带的设计形式统一，确保孩子绝对安全。高脚椅稳固而轻便，可以被稍大点儿的孩子轻松移动。我们的产品不仅关心孩子的成长，也同时关注他们的生活和发展。

The Leander chair is comfortably springy. The chair is adjusted to child's age and needs. The depth of the safety brace is adjusted — without using tools. The Leander chair is also a comfortable adult chair.

Children love to move and sitting quietly at the table can be a challenge for the little ones.

The designer wanted to make a chair that provided the possibility for movement, which was simultaneously comfortable to sit on. The designer found the solution in moulded wood, which is a friendly and flexible material.

The result was a living piece of furniture that bounces slightly when the child moves in the chair and it gives a pleasant and calming sensation.

The designer has chosen soft and organic forms, which are aesthetically related to the other pieces of Leander furniture — the forms are consistent in the back and brace, which securely embrace the child.

The chair is strong and stable and yet does not weigh very much so the slightly older child can move the chair around. The designer wanted to create a chair for the life and development of the child — and not just for growth.

Leander Changing Table

Leander

我们的目标是要让可变桌与 Leander 的其他产品相得益彰，而不是互抢风头。

用户若有需要，可以将底部书架或抽屉移开，简单组装后可变桌就可以轻松变身为写字台，桌高可适当调节，标准高度为 73 厘米。

高侧一端形成封闭空间，便于集中精力工作。桌面有圆弧形切口，方便电源线从此处通到桌后，以保持外观整齐。

这样，可变桌在不占用卧室过多游乐空间的情况下就能满足孩子的学习需要。成人亦可将这种时尚电脑桌放置在客厅或餐厅。

The aim of the changing table was to create a piece of furniture that supplements the design of the other pieces of Leander furniture. They must harmonise instead of battling for attention and attempting to overshadow each other.

With a quick restructuring, the changing table can be converted into a desk/workstation. Remove the bottom shelf and, if needed, the drawer. The tabletop can be set at the desired height. Standard desk height is about 73 cm.

The tall sides create an enclosed space, contributing the peace required to concentrate on your work — a modern version of a traditional desk. A discrete arc is cut in the tabletop, providing a hole for running cables down the back of the table. This keeps them out of sight and preserves the overall attractive appearance.

The changing table thus becomes a desk that meets your child's needs during the first years of school, when it shouldn't take up too much valuable play space in the bedroom. For adults, it provides an extra computer desk in the living room or kitchen-dining area that is discrete and stylish.

Index

索　引

Magis
www.magismetoo.com

modernconvenience
www.modernconvenience.com

modmomfurniture
www.modmomfurniture.com

MOOOI
www.moooi.com

Nurseryworks
www.nurseryworks.net

nikazupanc
www.nikazupanc.com

ninetonine
www.ninetonine.es

notNeutral
www.notNeutral.com

peterbristol
www.peterbristol.net

PLAY+
www.playpiu.com

Playsam
www.playsam.com

plons
www.plons.info

prodiz
www.prodiz.pl

sirch
www.sirch.de

Scoops Design
http://scoopsdesign.bigcartel.com/

Straightline Design
http://straightlinedesigns.wordpress.com/

Studio Laurens van Wieringen
www.laurensvanwieringen.nl

Studio Makkink & Bey BV
http://www.studiomakkinkbey.nl/

smarin
www.smarin.net

tau
www.tau.de

Vitra
www.vitra.com

W I H I L
www.purflo.com

Wishbone Design Studio
www.wishbonedesign.com

ASAI
www.asai-hagou.com

if-j
www.if-j.org

IO KIDSDESIGN
www.iokidsdesign.co.uk

后 记

本书在翻译过程中得到了曲乐的大力帮助，在此表示感谢！